Houcem Sammari
Monia Trabelsi

Lipid profiles of Symphodus tinca from the Tunisian coast

AF308482

Houcem Sammari
Monia Trabelsi

Lipid profiles of Symphodus tinca from the Tunisian coast

Comparative study

ScienciaScripts

Imprint

Any brand names and product names mentioned in this book are subject to trademark, brand or patent protection and are trademarks or registered trademarks of their respective holders. The use of brand names, product names, common names, trade names, product descriptions etc. even without a particular marking in this work is in no way to be construed to mean that such names may be regarded as unrestricted in respect of trademark and brand protection legislation and could thus be used by anyone.

Cover image: www.ingimage.com

This book is a translation from the original published under ISBN 978-620-6-71776-8.

Publisher:
Sciencia Scripts
is a trademark of
Dodo Books Indian Ocean Ltd. and OmniScriptum S.R.L publishing group

120 High Road, East Finchley, London, N2 9ED, United Kingdom
Str. Armeneasca 28/1, office 1, Chisinau MD-2012, Republic of Moldova, Europe
Printed at: see last page
ISBN: 978-620-7-94224-4

Copyright © Houcem Sammari, Monia Trabelsi
Copyright © 2024 Dodo Books Indian Ocean Ltd. and OmniScriptum S.R.L publishing group

Contents

Introduction.. 6

CHAPTER I: Literature review.. 8

I-1-Systematic 9

1-1-Synonyms 10

1-2-Non vernacular... 10

I-2-Geographical distribution... 11

I-3-Species discrimination criteria 11

3-1-Morphological criteria... 11

3-2-Coloring.. 13

3-2-1- Females 13

3-2-2- Young males 13

3-2-3- Adult males... 14

1-4-Healing 16

I-5-Reproduction 16

5-1-Optional successive protogynous hermaphroditism................................... 16

5-2-Laying period 16

I-7-Dietary regime 17

I-8-Biotope... 18

II-Lipid composition.. 18

II-1-Nutritive values of fish 18

II-2-Fatty acid classification and nomenclature 19

2-1- Structure of fatty acids 20

2-2-Saturated fatty acids 20

2-3-Monounsaturated fatty acids.. 21

2-4-Polyunsaturated fatty acids.. 22

III-Fish lipids.. 23

III-1-Lean fish 23

III-2-Oily fish 23

III-3-Semi-fatty fish 24

III-4-Action of external factors on lipid content 24

4-1-Power supply 24

4-1-1-Excess dietary lipids in fish... 25

4-1-2- Aquaculture 25

4-2-Temperature 25

4-3- Age 26

4-4- Sexual maturity 26

III-5-PUFAs in fish 27

5-1-Physiological roles of PUFAs in fish 27

5-2-Distribution of different types of fatty acids in cellular triglycerides 28

III-6-Lipid metabolism in fish 29

III-7-Metabolic pathways involved in the formation of lipid deposits 30

III-8-Fatty acid neo synthesis 30

III-10-Oxidation of lipids 30

IV-Benefits of PUFA consumption 31

IV-1-Effects of PUFAs on the cardiovascular system 31

IV-2-Hepatic steatosis 31

IV-3-Age-related macular degeneration 32

IV-4-AGPI and depression 32

IV-5-AGPI and schizophrenia 32

IV-6-Stress and aggressiveness 32

IV-7-Asthma and bronchoconstriction 33

IV-8-Adult, mother and newborn 33

Materials and methods 35

I-Sampling 35

I-1-Biological material 35

I-2-Precautions and storage 36

II-Lipid composition 36

II-1-Total lipid extraction 36

1-1-Principle of the method 36

1-2-Extraction 36

II-Methylation 37

II-1-Obtaining fatty acid methyl esters 37

II-2-Experimental protocol 38

III-Fatty acid separation 38

IV-Identification of fatty acids 39

Results 41

I-Comparisons of hepatho-somatic and gonado-somatic ratios between males and females of *Symphodus tinca* at the three stations (month of March) 41

II-Quantitative study of tissue lipids 42

II-1- Comparison of total lipid content of *Symphodus tinca* from the marine population as a function of sex (month of March) 42

II-2-Comparison of lipid profiles of *Symphodus tinca* from the island population as a function of sex (month of March) 42

II-3-Comparison of lipid profiles of male and female *Symphodus tinca* from the lagoon station (month of March) 43

II-4-Comparison of lipid profiles of *Symphodus tinca* females in three environments (month of March) 44

II-5-Comparison of lipid profiles between Symphodus tinca males from three environments (month of March) 45

III-Qualitative analysis of fatty acids 46

III-1-Comparison of fatty acid composition between males and females of the *Symphodus tinca* marine population (month of March).. 46

1-1-Comparison of flesh fatty acid composition between individuals ($\male$+$\female$) of the marine population 46

1-2-Comparison of the fatty acid composition of the livers of individuals ($\male$+$\female$) from the marine population. 47

1-3-Comparison of gonadal fatty acid composition of individuals ($\male$+$\female$) from the marine population. 48

1-4-Comparison of major fatty acids of individuals ($\male$+$\female$) from the marine population....... 49

III-2-Comparison of fatty acid composition of males and females from the *Symphodus tinca* lagoon population (month of March)... 51

2-2-Comparison of the fatty acid composition of the livers of individuals ($\male$+$\female$) from the lagoon population. 52

2-3-Comparison of gonadal fatty acid composition of individuals ($\male$+$\female$) from the lagoon population. 53

2-4-Comparison of major fatty acids of individuals ($\male$+$\female$)from the lagoon population. 54

III-3-Comparison of fatty acid composition of males and females from the *Symphodus tinca* island population (month of March)... 56

3-1-Comparison of the fatty acid composition of the flesh of individuals ($\male$+$\female$) from the island population. 57

3-2-Comparison of the fatty acid composition of the livers of individuals ($\male$+$\female$) from the island population. 58

3-3-Comparison of gonadal fatty acid composition of individuals ($\male$+$\female$) from the island population. 59

3-4-Comparison of major fatty acids of individuals ($\male$+$\female$) from the island population. 61

III-4-Study of variations in omega-3 and omega-6 content (month of March)..................... 63

4-1-Study of the omega-3 and omega-6 composition of the *Symphodus tinca* marine population 63

4-2-Comparison of the omega-3 and omega-6 composition of the *Symphodus tinca* population occupying lagoon waters 64

4-3-Comparison of omega-3 and omega-6 composition between males and females of *Symphodus tinca* on island coasts (month of March) 65

III-5-Study of variations in the ω6 ω3 ratio in the three organs of the three populations 66

Discussion 69

Conclusion 72

Bibliographical references 73

Appendices 90

Summary

Symphodus tinca L. 1758, commonly known as the peacock wrasse, is a medium-sized benthic osteichthyan fish with an important ecological role due to its ability to withstand wide variations in environmental physicochemical parameters. S. *tinca* is found in various marine, lagoon and island aquatic environments, with a biomass that has been declining over the last three decades.

This study compares the lipid profiles of a selected species belonging to the Labridae family: *Symphodus tinca*. Captured in different aquatic environments: the marine environment of Tabarka, the lagoon environment of Ghar Elmelh and the island environment of Djerba.

Firstly, the lipid contents of the flesh (Ch), liver (F) and gonads (G) were determined in *Symphodus tinca,* showing that lipid accumulation occurs primarily in the liver, secondarily in the gonads and finally in the muscles.

By comparing the biochemical composition of the fatty acids of these three organs in the three populations studied, we confirm that there is a variation in SFA, MUFA, PUFA, ω6 and ω3 levels according to sex and the environment populated, with speciation of the fatty acid arrangement for each organ.

These variations in lipid profile are essentially due to a combination of biotic and abiotic factors.

Key words: lipid profile, biochemical composition, SFA, MUFA, PUFA, ω6, ω3, *Symphodustinca,* Labridae, Tabarka marine environment, Ghar Elmelh lagoon environment and Djerba island environment.

Introduction

The Mediterranean Sea covers an area of around 2.5 million km2. It communicates with the Atlantic Ocean via the Strait of Gibraltar and with the Indian Ocean via the Suez Canal. It is also linked to the Black Sea by the Sea of Marmara.

From a hydrological point of view, the Mediterranean is generally characterized by a negative water balance and highly seasonal surface temperatures. In particular, its waters are not very productive, due to a lack of nutrient salts, especially as one moves away from the Strait of Gibraltar.

Another important oceanographic feature of the Mediterranean is that there are two main basins, an eastern and a western one, separated by the Strait of Sicily.

These main hydrological and oceanographic characteristics have an impact on the biology and ecology of fishery resources, as each of these 2 basins is ecologically made up of different units (Fonteneau, 1995).

The fish fauna of the Mediterranean, for example, is highly diversified: there are currently over 600 species of marine fish, most of which originate from the Atlantic (Quignard and Tomasini, 2000). However, due to the East-West gradient in temperature and salinity, the eastern Mediterranean is home to almost 400 species. The three main groups of marine fish present in the Mediterranean Sea are :

Agnatha, or jawless fish, which include lampreys and hagfish. These species are rarely found in the Mediterranean Sea, and are virtually absent from the eastern Mediterranean (Abdul Malak et al., 2011).

Chondrichthyans, cartilaginous fishes, are represented by 76 native species of sharks and rays in the Mediterranean Sea. They are distinguished by their primitive cartilaginous skeleton (Cailliet et al., 2005; Camchi et al., 1988).

Osteichthyans, or teleosts, or bony fish, characterized by their bony skeleton. They include some 442 species native to the Mediterranean (Abdul Malak et al., 2011).

Despite the considerable economic importance of these fish in Tunisia, crenilabra are widespread teleosts: they are caught from the north to the extreme south of the country. Although their size does not present any disadvantages for their marketing: their flesh has no particular taste, so they are of little economic value in the local fish market.

Fat is one of the major determinants of overall fish quality. Their content, composition and distribution in the different body compartments of fish are currently the subject of abundant literature, and even define a classification hypothesized by Ackman (1995). It therefore seemed interesting to carry out a review of fatty acids by analyzing the lipid composition of a benthic fish belonging to the Labridae family: The rattlesnake *Symphodus tinca.*

The main aim of the present study is to highlight and enhance the nutritional quality of peacock kelp. This can only be achieved by studying the fatty acid content and composition in three different organs involved in lipid metabolism. To this end, we set up a comparative analysis of the fatty acid composition of three populations from the Tunisian coast and during the same month, with the aim of detecting intra- and interstationary variations.

First, we characterized 3 different organs involved in lipid metabolism by observing the corresponding total lipid levels and fatty acid composition in each organ.

Secondly, we wanted to determine the variations in lipid content and fatty acid composition for the 3 deposition sites (flesh, liver and gonads), as a function of individual sex and station.

The aim of this experiment was to demonstrate the possible specificity of each tissue (lipid content and fatty acid composition).

Finally, we detail some explanatory approaches to these fluctuations affecting the quantity and quality of incorporated lipids, with reference to several biotic and abiotic factors.

CHAPTER I: Literature review

Labridae: perciform teleost fish belonging to the Osteichthyes, the most diverse superclass of living fish.

This family is represented by 60 genera with a total of 500 species (Nelson, 1994). The fossil history of labrids dates back to the Tertiary and Lower Paleocene (Berg, 1958). They are widely distributed, inhabiting the Atlantic, Indian and Pacific Oceans.

Generally occupying warm, tropical marine waters (Fisher et al., 1987). Several species are also found in the southern and eastern Mediterranean, with 20 species cited (Quignard, 1978), 14 of which have been recorded in Tunisian waters.

These fish come in a wide variety of shapes, sizes and brilliant colors, most with impressive liveries. Some are small (less than a dozen cm), while others, like the Napoleon, can reach several meters in length.

They are essentially coastal, occupying a bathymetry ranging from 0 to 80 meters in depth (Quignard and Pras, 1986). Their diversity is greatest in the tropics, and decreases towards higher latitudes. Coral reefs are found on several types of seabed: rocky seabed (Fisher et al., 1987), rocky seabed and bare sand (Fulkiger et al., 1981), but they often fill in seaweed beds, lawns and phanerogams (Casabianca et al., 1972-1973; Augier et Bouderesque, 1979; Harmelin-Viven, 1982; Quignard et Zaouali, 1980). Also coral reefs (Rivaton, 1989; LeTourneur, 1991), but moderately silty bottoms (Fisher et al., 1987).

The trophic spectrum of labrids is very much based on benthic invertebrates, and is essentially made up of crustaceans (shrimps), decapods, isopods and small gastropod molluscs.

However, for those in the Gulf of Lyons, their diet consists mainly of echinoderms and molluscs (Quignard, 1966). Similarly, other species are piscivorous, herbivorous (Fisher et al., 1987) and notably omnivorous according to (Thresher, 1983). These authors based their studies on the stomach contents of fish, a euryphagous family with fairly strong teeth that are completely fused with the lower pharyngeal bone, making the teeth a veritable pebble. Some use their snouts to turn over stones and pieces of coral to filter out hidden invertebrates (Moyle and Cech, 2000).

Most labrids rely on vision to find their prey, and their activity is diurnal (Fisher et al., 1987). Their diet shows such variation according to species and season (Quignard,

1966), that their hue ensures excellent homochromy with the seabed. This coloration can change at night, as has been proven in some marine species.

In times of danger, they burrow into the phanerogam meadows between small stones and cracks (Bell and Harmelin-Viven, 1982; Harmelin-Viven, 1982).

As far as reproduction is concerned, they are all Oviparous. The slope may be pelagic or demersal; in the 2ème case, the males prepare nests to protect the eggs inside. This nest is beaten into fragments of sea grass (Soljan, 1930 and 1931). The nest is protected by the male, who remains at the opening.

A phenomenon of facultative, protogynous hermaphroditism has been described, affecting the majority of individuals in populations (Ross et al., 1983). During the breeding season, they are gregarious, but outside this period, they are solitary (Hoffman et al., 1985), with intra- or interspecific aggressiveness.

Like most members of the large Labridae family, the Peacock Rattlesnake, a rock fish par excellence, lives in rocky areas most of the time. It can be seen passing through open areas, but is most often to be found in stones or in grass beds, always in the immediate vicinity of the faults. This is where it will hide in case of danger. A homebody on its seasonal outfits, the kestrel emerges from its hole during the day to feed in the immediate vicinity of its nest.

In the presence of troublemakers (hunters, bathers...), our fish hides and remains in the hole until the intruders leave. Like most rock fish, *Symphodus tinca* is a diurnal fish, fished only between sunrise and sunset (Quignard, 1966).

They showed a certain homochromy with their environment, i.e. they could change color to camouflage themselves. This hue transition is predicated on stress: in the event of fear, danger or threat, or for males dominated by others, the coloration is much paler.

Young individuals often travel in small shoals of a few individuals accompanied by an adult female (Le Bris et al., 2013).

To build his nest, the male chooses a small cave (usually a large depression in the rock face) at a depth of 10 to 15 m, lining the vertical wall with algae collected from the surrounding area (Le Bris et al., 2013).

I-1-Systematic

In biology, classical classification refers to the traditional scientific classification based on a comparative analysis of the morphological characteristics of species. The Swedish botanist Carl Von Linné (1707-1778) was its initiator, attempting to apply it to all the living beings he knew. According to Linnaean classification, *Symphodus tinca* occupies the following systematic position:

Kingdom: Animal.

Subregion: Metazoans.

Branch: Ropes.

Subdivision: Vertebrates.

Superclass: Osteichthyes.

Class: Actinopterygians.

Subclass: Neopterygians

Infra-class: Teleosts.

Superorder: Acanthopterygians.

Order: Perciformes.

Suborder: Labroidae.

Family: Labridae.

Genus: *Symphodus.*

Species: *tinca.*

1-1-Synonyms

According to these authors, *Symphodus tinca* is mentioned by various synonyms:

Labrus tinca, linnaeus, 1758.

Crenilabrus pavo, Brunnich,1768.

Labrus lapina, Frosskal ,1775.

Lutjanus geofroyensis, Risso,1810.

Labrus polychrous, Pallas, 1831.

1-2-Non vernacular

Given its vast geographical range, *Symphodus tinca* has several names, which vary from country to country:

Tunisia: khodhir, Soltane, Lapse.

Algeria: Sabonero.

Italy: Laggion

France: Crénilabre peacock, Crénilabre slice.

Marseille: Roucaous.

Monaco:Ruchétanca.

Spain: Peto.

English:Peacockwrasse.

I-2-Geographical distribution

The geographical range of *Symphodus (Crelinabrus) tinca* extends from the eastern6 Atlantic: from Morocco to the northern coasts of Spain (Quignard and Pars, 1986), crossing the entire Mediterranean and as far as the Black Sea (Fisher et al., 1987). This area of 44 N-21 N, 18 W-42 E has a subtropical climate (figure 1).

Figure 1: Geographical distribution of *Symphodus tinca.* Source (http://www.fishbase.org)

I-3-Species discrimination criteria

3-1-Morphological criteria

The peacock wrasse is a medium-sized fish, ranging in length from 10 to 25 cm, with a maximum of 44 cm (Fisher et al., 1987). According to Bradai (2000), it is one of the largest wrasses on our coasts and the most common in the Mediterranean. Males can reach 40 cm, while females do not exceed 25 cm.

Figure 2 shows the morphology of the labre:

- The body is ovoid, slightly compressed, a massive fish with an elongated body (Le Bris et al., 2012).

- The head is elongated and longer than the body (Fisher et al., 1987).

The snout is pointed and longer than or equal to the pre-orbital space, with a small mouth featuring thick lips with 6 to 9 folds, and numerous cephalic pores (Fisher et al., 1987).

- The teeth are strong, sharp and caniniform, arranged in a single row on each of the 2 jaws (Bouchot and Pras, 1987). The teeth are made up of 2 parts, always short:

One is horizontal, the other descending.

The premaxilla also has 2 branches, one horizontal and the other ascending.

Juveniles have a serrated edge on the lower part of the operculum, which disappears with age.

- There is a small dark spot on the caudal peduncle, but sometimes 5 small black spots on the dorsal and 3 on the anal (Bauchot and Pars, 1987).

- Thin, cycloid scales cover the entire body except for the snout and the space between the eye sockets. They are elongated, medium to large in size, increase in size with age, and are covered by a tegumentary fold.

- The lateral line has 33 to 38 scales (Quignrad, 1966; Tortonese, 1975). This line is continuous, surmounted by 3 and 1/2 or 4 and 1/2 rows of scales and dips and curves toward the base of the body near the caudal peduncle.

- The scales on the cheeks are ornamented in 4 to 6 lines.

- On the first branchial arch, the number of branchiospines varies from 13 to 16 (Tortonese, 1975).

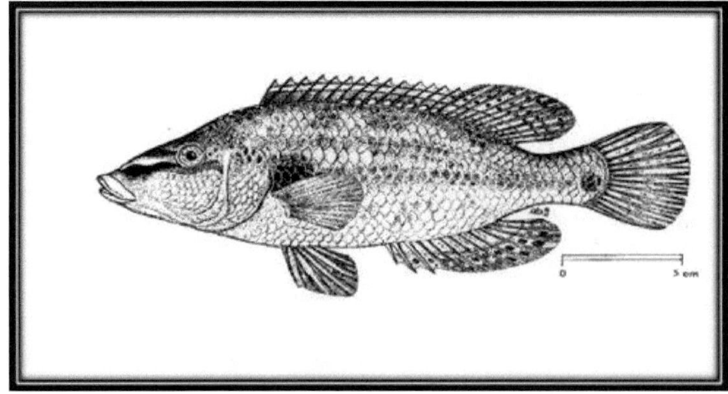

Figure 2: *Symphodus (Crenilabrus) tinca* (linnaeus, 1758).

3-2-Coloring

This species displays a sexual dichromatism that becomes all too apparent during the breeding season. Thus, (Ouannes-Ghorbel, 1996) distinguished 3 liveries with greater or lesser differences in pigmentation.

3-2-1- Females

At different age classes, females are always light to dark brown, slightly greenish (Figure 3). The belly is whitish to silvery, and the pectoral and pelvic fins are whitish to yellowish.

The anal and dorsal fins are light brown with dark pigmentation. The caudal peduncle has a dark spot that is present on all individuals, whether male or female. The back is consistently dark brown, with a grayish to greenish tinge at the front. Cheeks and lips are light gray to silvery.

Figure 3: Female *Symphodus tinca* "below" (Le Bris, 2007).

3-2-2- Young males

Young males are similar in color to females (Figure 3). They are paler than adult males and rather blander. Their hue is light brown to reddish.

Figure 4:Peacock rattlebill: young male (Le Bris, 2006)

3-2-3- Adult males

They are more colorful and dazzling than females, with golden to yellowish bellies, greenish pelvic fins and yellow pectorals.

The other fins are blue and yellowish to greenish at the base, tinged with blue or red at the tip (Figure 5).

Iris color varies from yellow to green. As in females, the urogenital papillae are small and white or yellow, the cheeks are yellow, and in front of the eye there is a blue spot. Lips are yellow or green. The sides and back are dark olive green with yellow highlights.

During the cooler months, males become brighter (Figure 6). They take on a composite hue of blue, green or yellowish-green, the red spots arranged in 4 to 5 lines on either side of the lateral line become brighter, and the part above the head turns blue.

Figure 5: Adult male *Symphodus tinca* in mating season (Le Bris, 2004).

Figure 6: Adult male crenilabre slice not in breeding season (Le Bris, 2007).

1-4-Healing

Some individuals of this wrasse had quite singular areas of scales in different body regions, reinforcing the existence of a regeneration phenomenon (Vincent, 2009).

According to (Quignard, 1966), fish have an incredible healing capacity, capable of regenerating huge wounds or completely torn-off fins. That's why you rarely come across a fish with an amputated fin.

This is obviously of interest to human medical research: there are many publications on the subject, particularly in zebrafish.

I-5-Reproduction

5-1-Optional successive protogynous hermaphroditism

This species exhibits successive and facultative protogynous hermaphroditism (Quignard and Pras, 1986). This has been demonstrated in the Gulf of Gabes (Ouannes-Ghorbel, 1996 and 2003).

(Pealaoro and Jards, 2003) have shown that the sex ratio is in favor of males, who dominate almost all age classes, and that from 29 cm upwards, all individuals are male.

(Ben Slama et al., 2010) measured the sex ratio from September 2004 to August 2005 of crenilabres from the northeast coast of Tunisia, which was in favor of males (1.38:1), and individuals larger than 27 cm were predominantly male, supporting the hypothesis of hermaphroditism in this species. In the post-spawning period, males outnumber females.

Indeed, facultative protogynous hermaphroditism is verified by histological sections, with some specimens showing a mixed gonad. The histological study of these gonads made it possible to describe the evolution of hermaphroditic gonads. These glands showed simultaneous islands of degenerating male and female germ cells.

Secondary males, resulting from the sexual inversion of females, are said to have lamellar testes (Ben Slama et al., 2010).

5-2-Laying period

In the southern Mediterranean, crénilabre spawning extends from April to June (Ouannes-Ghorbel et al., 2002). In the northern Mediterranean, all along the French coast and in the Gulf of Lyon, this period is also the same as that of the Tunisian coast.

It extends from April to June, with a possible extension into early July (Quignard, 1966). This is also described by Dieuzeide (1955), on the Algerian coast. According to (Ouannes-Ghorbelet et al., 2002), this shift is attributed to photoperiod rather than water temperature.

Peacock wrasse reproduction takes place in 3 phases:

- Pre-pontal or maturation phase characterized by 3 periods: a period of slow development of the gonads, followed by a period of maturation or more rapid development.

- Gamete emission phase, where there is a drop in this ratio, from April to June, during which the weight of the gonads falls enormously.

- Sexual rest phase.

I-7-Dietary regime

Given its biotope, the peacock wrasse feeds on shallow benthic organisms: crustaceans (isopods and shrimps) and gastropod molluscs are its main prey. These trophic preferences are gender-neutral and almost identical for all age classes. As age increases, the percentage of isopods consumed decreases in favor of shrimp and other large molluscs (Ouannes-Ghorbel and Bouain, 2006). This is explained by the fact that animal growth is linearly related to mouth width development (Ross et al., 1978; Stoner, 1980).

Apart from this trophic ontogeny, the diet of slice crenilabers shows slight seasonal changes in the quantity and quality of their trophic spectrum, but remains based on crustaceans and mollusks.

Nevertheless, during the winter period, the stomach contents of adults show greater consumption and present other prey, but in small proportions: (amphipods, polychaetes, algae, cephalopods, foraminifera, small teleost fish, bivalves, diatoms) (Ouannes-Ghorbel, 2003).

Consumption of these prey items may be accidental, as in the case of algae, since wrasses feed at high intensity in winter to optimize the necessary energy expended during gonad maturation (Ouannes-Ghorbelet et al, 2002).

These fish hardly feed at all during the cool season. The stomach contents of the specimens were analyzed. Of the total, (87.4) of the digestive tracts were empty. This percentage varies over the year, with a maximum during the cool period. (Quignard, 1966) also confirmed that the trophic spectrum of *Symphodus tinca* adults from the south French coast is vast. He isolated amphipods, foraminifera, decapods, annelids and amphineurs from the stomachs of small fish, with the exception of crustaceans and molluscs.

This change in feeding behavior, according to (Ware, 1972; Stoner and Lingviston, 1984), is correlated with the increase in mouth size. It enables fish to capture a wider range of prey sizes.

I-8-Biotope

Symphodus tinca is a coastal species that frequents rocky bottoms, seaweed and Posidonia meadows, phanerogams and seagrass meadows (Selourde and Chauvet, 1986; Quignard and Zouali, 1980), and coral reefs (Le Tourneur, 1991). They occupy a bathymetry of 80 meters (Fisher et al., 1987), limited to 40 meters in the Gulf of Gabes (Ben Othman, 1973).

They can enter brackish lagoons (Bourquard, 1985), unlike adults, juveniles occupy shallow depths (Tortonese, 1975). In Tunisia, this species is found in northeastern and southern regions (Azouz, 1974).

II-Lipid composition

II-1-Nutritive values of fish

From a nutritional point of view, fish flesh is rich in highly digestible proteins, minerals (notably phosphorus, calcium, magnesium and iodine), vitamins (particularly vitamin D and other fat-soluble vitamins) and fatty acids (mono- and polyunsaturated n-3 series, such as eicosapentaenoic acid (EPA) and docohexaenoic acid (DHA)).

In fact, it represents an excellent foodstuff with nutritional characteristics that are unique among animal products, and at an affordable cost. This explains why fish is such an important part of the human diet. Worldwide per capita fish consumption has risen over the past four decades, from 9 kg per person per year in 1961 to 16.5 kg in 2003 (FAO, 2006).

Among its various biochemical components, lipids are among the major determinants of overall fish quality. Their content, composition and distribution in the different body compartments of fish even define a classification hypothesized by Ackman (1966).

However, fatty acids are currently the subject of abundant literature. Polyunsaturated", "omega 3" and "conjugated linoleic" are all terms with positive connotations used in human health.

Numerous studies show that increasing attention is being paid to "good" and "bad" sources of fatty acids for humans. These include animal products, particularly seafood, especially fish.

It is therefore important to be able to qualify and quantify the fatty acids present in these products. Given the popularity of these compounds, which are attributed with numerous beneficial physiological effects, they are used as raw materials in food supplements, medicines and cosmetics. They are used as raw materials in food supplements, medicines and cosmetics (Veylon, 1977; Hattori et al., 1998; Holmström and Kjelleberg, 1999; Hellio et al., 2000).

Like other metabolites (sugars, proteins, etc.), lipids are of potential interest in all these fields, and particularly in the area of biological activity. For example, Watkins et al (2001) provide an overview of the activities of lipids and fatty acids on the biology and cellular functions of bone. These features, and in particular the richness of PUFAs, make the marine environment an interesting source of bioactive lipids.

II-2-Fatty acid classification and nomenclature

Fatty acids are part of the lipid family of water-insoluble organic molecules. Lipids have been the subject of numerous classifications. Hennen (1995), for example, classified these molecules into 6 categories: triglycerides, glycerophospholipids, sphingolipids, terpenoids, sterols and steroids, and fatty acids.

Fatty acids are the basic constituents of saponifiable lipids, whether simple or complex. Rarely, they are present in the non-esterified AGNE state. All fatty acids are made up of a hydrocarbon chain, which may be linear or branched, with an even number of carbons ranging from 4 to over 30, and in the marine environment from 14 to 24. Depending on the length of their chain, these compounds are divided into 3 groups, a more planned classification of which is proposed by Canler (2001) (Table 1).

Table 1: Classification of fatty acids by carbon number (Canler, 2001)

Number of atoms of carbon	Chain length	Common name
6 à10	Short or medium chain	Butyric acids

| 12 à22 | Long chain | Fatty acids |
| >22 | Very long chain | Waxy acids |

2-1- Structure of fatty acids

In functional terms, these organic molecules present :

- A methyl group (-CH3) at one end.

- A carboxyl group (-COOH) at the other end gives the molecule its acidic character. Another discriminating criterion for fatty acids is the number and location of the double bond (unsaturation) between the carbon atoms. Thus, there are three types of fatty acids: saturated fatty acids, which have no double bonds; monounsaturated fatty acids, which have one double bond; and polyunsaturated fatty acids, which have two to six double bonds, usually separated by a methylene group (CH2), (Budge et al., 2006). Fatty acids are named according to the notation A : Bn-X where A is the number of carbons, B is the number of double bonds and X is the position of the first double bond relative to the terminal methyl group.

2-2-Saturated fatty acids

These are fatty acids in which all the carbon atoms are saturated with hydrogen, so that each carbon atom carries the maximum number of hydrogen atoms, and no further hydrogen atoms can be added to the molecule.

Thus, all bonds between carbons are single and there are no double bonds (unsaturation). The general chemical formula for saturated fatty acids is :

CH3 - (CH2) n - COOH

Where: N = n+2

N: total number of carbon atoms

They are represented by a continuous series of fatty acids with an even number of carbons (4 to over 30 carbons) (Appendix 1). They were isolated from animal, plant and microbial lipids.

In fact, these saturated GAs can no longer be considered as a whole, as they differ in structure, metabolism and cellular functions (Hughes et al., 1996; Legrand and Rioux, 2010).

Palmitic acid (C16:0) and stearic acid (C18:0), mainly from animal products, are the main contributors to our diet. Numerous epidemiological studies have demonstrated the harmful effect of excessive SFA consumption on the development of cardiovascular disease.

That's why we recommend limiting consumption of fats with high levels of SFAs.

This is the case with butter, cream, beef tallow, cocoa butter, copra, palm and palm kernel oils, certain margarines and certain prepared foods, which may contain high levels of these substances (see nutritional labeling).

In addition to this external supply, their endogenous origin is highly varied: palmitic acid (C16:0) is the most intensively synthesized as the first fatty acid produced from glucose and acetate, in a pathway that goes directly from 2 to 16 carbons.

Most naturally occurring fatty acids have an even number of carbons and a linear chain, but there are a few with an odd number of carbons and a branched chain. For example, isovaleric acid is formed when a methyl group is attached to the penultimate carbon atom.

2-3-Monounsaturated fatty acids

They account for more than half of the fatty acids found in plants and animals:

- A single double bond: monounsaturated acids.

In monounsaturated fatty acids, the position of the single double bond can be expressed as :

- Either starting from the carboxyl (1st carbon); the symbol is Δ.

or starting from methyl (last carbon); the symbol is omega ω. In clinical medicine and biology, the most common designation for unsaturated fatty acids is omega (ω) (Appendix 1).

Monounsaturated fatty acids (MUFAs) comprise two families, n-7 and n-9, of which oleic acid (C18:1, n-9 or ω9) is the most widespread. The general chemical formula for monounsaturated fatty acids is :

$$CH3 - (CH2) x - CH = CH - (CH2) y -COOH$$

With : N = x+y+2.

X: position of unsaturation starting from the methyl group.

Y: position of unsaturation starting from the carboxyl group.

Like saturated fatty acids, monounsaturated fatty acids (MUFAs) come from both endogenous synthesis (in humans, as in virtually all living beings) and from the diet. They are present in olives, rapeseed, nuts (pistachios, almonds, hazelnuts, cashews, pecans), peanuts, avocados and animal products.

Their endogenous synthesis is carried out by delta-9-desaturase, which introduces a double bond on palmitic acid and stearic acid, leading respectively to palmitoleic acid (C16:1 n-7) and oleic acid (C18:1 n-9).

2-4-Polyunsaturated fatty acids

Polyunsaturated fatty acids with less than 18 carbon atoms are absent or present in extremely low quantities in vegetable and animal fats, but C14 and C16 fatty acids have been mentioned in marine animal oils.

There are two families of polyunsaturated fatty acids (PUFAs): n-6 or omega 6 (ω6) and n-3 or omega 3 (ω3) (Appendix 1).

These fatty acids are derived respectively from linoleic acid (C18:2 n-6) and α-linoleic acid (C18:3 n-3). Omegas are polyunsaturated fatty acids whose first double bond is located on the third carbon from the terminal methyl group. Whereas omega-6s are polyunsaturated fatty acids whose first double bond is located between the sixth and seventh carbon from this G-terminus.

Long-chain polyunsaturated fatty acids are PUFAs with more than 4 double bonds, such as :

- C20:5 (5, 8, 11, 14, 17) (eicosapentaenoic acid).

- C22 : 5 (4, 8, 12, 15, 19) (docosapentaenoic acid).

- C22:6 (4, 7, 10, 13, 16, 19) (docosahexaenoic acid).

Figures 7 and 8 show the semi-developed formulas of EPA and DHA, the most abundant PUFAs in fish TAGs.

Figure 7: Semi-developed formula for Eicosapentaenoic acid

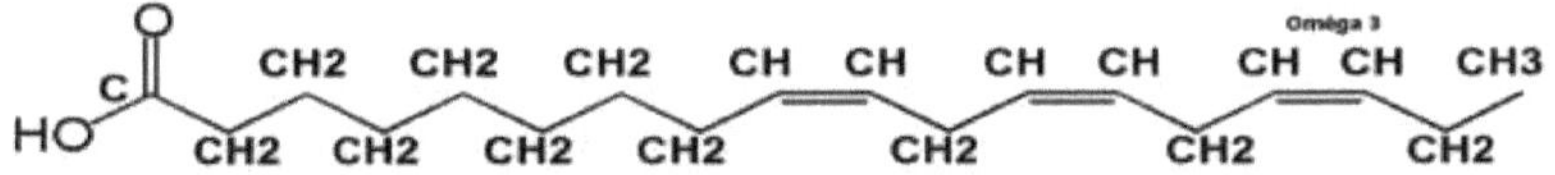

Figure 8: Semi-developed formula of Docosahexaenoic acid

These long-chain PUFAs are found in marine animal oils. C20 :5 ω3 and C22 :6 ω3 can be isolated from herring or cod liver oil, where their contents reach 9%, while C22 :5 (DHA) is found in almost all oils of marine origin (Sonntag, 1979; Aidos et al., 2002).

III-Fish lipids

Oils from different tissues are divided into 2 groups: phospholipids and triglycerides, of which the first group shows virtually no quantitative variation. Therefore, any decrease or increase in muscle fat is directly correlated with variations in triglyceride levels (Ando et al., 1993). The lipid content of fish flesh varies according to several factors, but essentially according to species. This criterion is the basis of a classification of fish invented by Ackman (1980) and Henderson et al. (1995):

III-1-Lean fish

These fish reserve almost all their lipids in the liver cells, with a high ratio that can range from 40% to 75% of liver weight. Lipids in the flesh can be as low as 1% or even less than 4%. These are essentially groundfish (hake, whiting, sole, scorpion fish, plaice, sea bream, skate and dogfish), considered to be lean fish.

III-2-Oily fish

These fish store their fat reserves mainly in muscle, to a lesser extent in skin tissue, and to a lesser extent in the abdominal cavity.
Their lipid content generally exceeds 15%. Pelagic fish include salmon, sprat, sardines, mackerel and herring. These fish have a high percentage of PUFAs and MUFAs.

III-3-Semi-fat fish

Also known as semi-fatty fish, their muscles contain 2 to 10% lipids. Their adipose deposits are essentially muscular and partially perivisceral (Swordfish, Tuna, Anchovy).

III-4-Action of external factors on lipid content

Apart from the interspecific and tissue differences mentioned above, the content and distribution of lipids in the different body compartments of fish varies according to environmental factors. Water temperature and salinity, as well as physiological factors such as age, sex, sexual maturation and nutritional background, have varying effects on lipids.

4-1-Power supply

Numerous studies have shown that the biochemical composition of fish can be modified quantitatively and qualitatively by dietary intake, in particular by diet composition and energy content (Shearer et al., 1997; Jobling et al., 1998; Johansson et al., 2000 and Médale et al., 2003). Although it is the absorbed energy content of the food that is decisive, there is a close relationship between lipid content of the food and body lipid content (Cowey, 1993; Kaushik, 1997).

In aquaculture, several studies have shown that an increase in the lipid content of the feed enables protein to be better utilized for growth. Lipids then serve as the preferred energy substrate (Takeuchi et al., 1978; Hillestad and Johnsen, 1994).

This is the notion of protein sparing by lipids, which not only improves growth performance, but also reduces nitrogen emissions by limiting amino acid oxidation (Kaushik and Oliva-Teles, 1985; Kaushik, 1997; Cho and Bureau, 2001; Hemre et al., 2002; Dabrowski and Guderley, 2002).

In freshwater fish, for example, feeding isoprotein diets with different lipid contents (ranging from 5% to 20% lipids) results in a significant increase in perivisceral lipid deposits and a moderate increase in muscle lipid content. Liver lipid content does not vary significantly (Takeuchi et al., 1978; Corraze and Kaushik, 1999).

These results have also been proven in other catadromous fish (salmon) (Hillestad and Johnsen, 1994).

In marine fish, such as sea bass (*Dicentrachus labrax*) or cod (*Gadus morhua*), liver lipids in particular increase in response to a high-fat diet. Muscle and visceral lipid contents are not significantly altered (; Nanton et al., 2001).

In fish, lipid digestibility often exceeds 92%, regardless of origin (animal or vegetable). The nature of the fatty acids making up lipids influences their digestibility, which increases with their degree of unsaturation (Takeuchi et al., 1979; Sigur Gisladottir et al., 1992; Johnsen et al., 2000). However, not all species have the same capacity to use lipids and spare proteins.

4-1-1-Excess dietary lipids in fish

High dietary lipid levels can lead to reduced digestibility, as observed in cod (Lie et al., 1988), catfish (Andrews et al., 1978) and sturgeon (Médale et al., 1991).
In turbot, a lipid content of over 10% in the feed leads to a slowdown in growth rate (Regost et al., 2001). Nevertheless, some species, such as rainbow trout, Atlantic salmon and halibut, tolerate high dietary lipid levels (up to 30% of dry weight in the feed) (Aksnes et al., 1996; Rasmussen et al., 2000; Tortensen et al., 2000).

4-1-2- Aquaculture

Over the past 25 years, the composition of commercial fish feeds has gradually evolved, with a decrease in protein content associated with an increase in lipid intake. In salmon, lipid content often exceeds 25% of the feed (Hemre and Sandnes, 1999), and can reach almost 40% of the feed (Einen and Roem, 1997; Torstensen et al., 2000). While these high-energy feeds improve growth performance, they also lead to an increase in lipid deposits (Watanabe, 1982; Corraze and Kaushik, 1999). Excessive development of these deposits can have negative repercussions on the quality of fish products.

4-2-Temperature

In particular, temperature is a major regulator of energy requirements and metabolic utilization of nutrients (Médale et al., 1991), as well as gastric filling rate and emptying time (Brett 1979; Kaushik, 1986). In general, nutrient absorption is slower in fish than in mammals, due to their lower body temperature. Absorption is diet-dependent, being slower in omnivores and faster in carnivores (Kapoor et al., 1975). At lower temperatures, fish may even stop feeding (Médale et al., 1991).

Thus, a rise in water temperature is accompanied by an increase in body lipid content, essentially resulting from an increase in the quantity of food ingested (Brauge et al., 1995).

Water temperature is also likely to affect dramatic fluctuations in the lipid composition of fish: an increase in PUFA content and a decrease in SFA content are observed as temperature decreases (Greene and Selivonchick, 1987), allowing fish to preserve the fluidity of membrane structures.

4-3- Age

Fish are distinguished from other vertebrates by their prolonged growth throughout their lives, under favorable environmental, physiological and dietary conditions (Greer-Walker, 1970; Stickland, 1983; Weatherley and Gill, 1987). So, necessarily, body lipid content tends to increase with increasing fish size (Denton and Yousef, 1976; Reinitz, 1983; Rasmussen and Ostenfeld, 2000).

This increase in lipid content is generally accompanied by a decrease in water content, while protein content remains relatively stable (Henderson and Tocher, 1987; Shearer, 1994; Jobling et al., 1998). This increase in body lipids with fish size mainly affects the preferred deposition sites of the species in question.

Thus, adipose tissue growth in adulthood is mainly due to an increase in the size of existing cells (hypertrophy), while the number of adipocytes is assumed to remain constant. Some studies (Faust et al., 1978; Faust and Miller, 1981) have shown that there may also be recruitment of new cells (hyperplasia) in response to high-fat or high-carbohydrate diets, but this phenomenon appears to be limited in mammals. In fish, different populations of spherical cells can be distinguished in adipose tissue, ranging in diameter from 5 to over 200 µm, with the majority measuring between 20 and 60 µm (Zhou et al., 1996).

The presence of large numbers of small adipocytes is characteristic of hyperplastic adipose tissue development. In fish, adipose tissue volume increases both by hypertrophy of existing cells and by hyperplasia, throughout the animal's life (Fauconneau et al., 1997; Gelineau et al., 2001).

4-4- Sexual maturity

Gonad development during sexual maturation leads to a mobilization of body lipids in fish. This mobilization of body lipids is more pronounced in females than in males, as the energy requirements for ovarian and oocyte development are higher (Henderson and Tocher, 1987). The storage sites affected by this mobilization differ according to reproductive stage.

III-5-PUFAs in fish

5-1-Physiological roles of PUFAs in fish

To better understand why fish are characterized by high levels of omega-3 and omega-6, which are not synthesized by their metabolic machinery. These levels depend on a number of factors, including diet and tissue physiology.

Fish are poikilothermic vertebrates, so their body temperature is determined by that of the surrounding environment. To escape and adapt to these sometimes acute thermal variations, they need to maintain their cellular plasticity.

This is solved by incorporating lipid structures into their tissues to withstand environmental variations.

These lipids are PUFAs (DHA and EPA), with long carbon chains and a high number of instantiations, giving the molecule enormous plasticity, which maintains cell membrane fluidity even at very low temperatures (Eldho et al., 2003). In some cases, DHA-rich lipids even form fluidity islands. This enables transmembrane proteins (Na+-K+ ATPase, cytochrome oxidases...) to assume the optimal conformation for their function (Arts and Kolher, 2009). Even at enormous temperature changes, the fluid mosaic retains all its functional capacities: absorption and vesicle formation, permeability, transport (Wassal et al., 2004). Other authors value the structural and modulating role of these essential fatty acids; Sargent et al. (1990) have shown that fish larvae need DHA to complete the development of the nerve tube, which is incomplete at hatching. At this stage, any deficiency or lack of DHA supply causes developmental arrest, SN disorders and, above all, ocular disorders. These nervous disorders have the following symptoms:

-Spectacular behavioral abnormalities in flight, predation and band building (migration).

- Pigmentation defects.

This tends to increase the lethality rate of larvae and juveniles in the wild, since they are vulnerable, slow and predator-prone.

Although PUFA-rich diets with inadequate DHA/EPA/ARA ratios can have remarkable effects on resistance to stress and immunity (Koven et al., 2003; Van Anholt et al., 2004; Arts and Koher, 2009).

APGI are also involved in the synthesis of various molecules such as prostaglandin and thromboxone (Schmitz and Ecker, 2008). They are involved in the regulation of various biological processes: vasoconstriction, immunity, hydromineral regulation (Arts and Kolher, 2009; Schmitz and Ecter, 2008).

Eicosanoids include EPA- and ARA-derived prostaglandins, which are antagonistic regulators (Olsen, 1998; Calder, 2009).

Eicosanoid levels are partially influenced by dietary intake (Olsen, 1998). This explains how the relative proportions of PUFAs in the diet can affect animal physiology.

From the point of view of trophic factors, in their natural environment, marine fish at the base of the trophic chain find these molecules in the phytoplankton they feed on, which have a high concentration of n-3 PUFAs (Sargent et al., 1989). These in turn serve as a source of n-3 PUFAs for fish at higher trophic levels (Watanabe, 1982; Reinitz, 1983; Henderson and Tocher, 1987).

On the other hand, the diet of freshwater fish is richer in linoleic and linolenic acids (Tocher, 2003). These fish must therefore convert these C18 fatty acids into PUFAs to preserve the fluidity of their membranes. Consequently, and unlike marine species, most freshwater fish have an excellent capacity to elongate and desaturate their C18 fatty acids into longer-chain fatty acids (Sargent et al., 1999).

5-2-Distribution of different types of fatty acids in cellular triglycerides

Given the nutritional impact of EPA and DHA, and the importance of PUFA bioavailability, it is essential to know exactly how these PUFAs are distributed on fish TAGs. To study this regiodistribution, Luddy et al (1963) invented an enzymatic method using pancreatic lipase. It hydrolyzes highly unsaturated GAs (over 20 carbons and 3 unsaturations) more slowly than GAs with 16 and 18 carbons, with little or no unsaturation (Entressrangles et al., 1961). GAs with 20:5 and 22:6 are found in the outer positions of fish oils.

The chemical method, using Grignard reagents, has been used to study various fish and marine mammal oils (Brocherhoff et al., 1963; Turon et al., 2002). These studies show that EPA and DHA, when derived from marine mammals, are most often located in external positions on TAG molecules. This difference in chemical structure is related to a physiological observation: the GA composition of the red blood cells of the Inuit (who in north-western Canada feed mainly on marine mammals) is very different from that of people who eat fish. The differences in GA regiocomposition in red blood cells could therefore be explained by their different diets.

III-6-Lipid metabolism in fish

The storage of fatty acids in reserve tissues is not the same in terms of quantity and quality, as demonstrated by (Pascal and Ackman, 1976). This indicates that the major role of fatty acids is to provide the energy required for all vital activities and, secondly, for tissue construction. Variation in the proportions of fatty acids accumulated and absorbed is directly correlated with oxidation.

With regard to synthesis, marine fish are unable to transform 18:2 (n-6) and 18:3 (n-3) into EPA and DHA. These 2 long-chain polyunsaturated fatty acids (omegas 3) must be supplied by dietary intake (Bell et al., 1986).

However, the synthesis of long-chain PUFAs (DHA, EPA, ARA) is energetically costly, and it is preferable to obtain them from the diet (Olsen, 1998).

Unlike freshwater fish, marine species with little or no $\Delta5$-desaturase activity are unable to carry out these bioconversions at any appreciable rate.

They must then recover long-chain PUFAs (DHA, EPA, ARA) via their food (Sargent et al., 1995; Sargent et al., 1999; Brett et al., 2009).

In contrast, freshwater fish possess the desaturases ($\Delta6$ and $\Delta5$) and elongases that enable them to synthesize all the PUFAs of the $\omega3$ and $\omega6$ series from the precursors 18 :3 $\omega3$ and 18 :2 $\omega6$ (Henderson and Tocher, 1987; Sargent et al., 1995).

Although these PUFAs are poorly synthesized, fish require permanent external supplies, so they are considered "essential" molecules or (Essential Fatty Acids: EFAs). Precise EFA requirements vary qualitatively and quantitatively, depending on the fish family considered (Sargent et al, 1999).

In fact, in the marine world, the primary producers of EPA and DHA are phytoplankton micro-algae. Fish produce few omega-3s. Farmed fish with no access to marine nutrients contain far less omega-3.

III-7-Metabolic pathways involved in the formation of lipid deposits

The main metabolic pathways involved in the genesis and degradation of lipid deposits in fish are generally comparable to those in mammals (Walton and Cowey, 1982):

- External dietary lipid intake.

- Endogenous synthesis of new lipids from other fatty or non-fatty nutrients.

- Blood and lymphatic transport and lipid uptake by various tissues.

- Lipid storage in reserve tissues.

- Lipolysis and the use of lipids for energy purposes.

III-8-Fatty acid synthesis

In fish, *de novo* synthesis of fatty acids is very limited in adipose tissue, and it is mainly the liver that synthesizes fatty acids from dietary carbohydrates and amino acids (Greene and Selivonchick, 1987). In fish, the main SFAs synthesized are C16:0 and C18:0, and to a lesser extent, C14:0; but their percentages vary according to species. The main fatty acids produced by fatty acid synthase are 16:0 and 14:0, in oily fish. In flatfish, they are 18:0 and 16:0 (Sargent et al., 1989).

Endogenous and trophically-derived SFAs can undergo elongation and desaturation to transform into other fatty acids.

Δ9-desaturase, or stearoyl-CoA desaturase (SCD), is the primary enzyme involved in the synthesis of monounsaturated fatty acids. It is inhibited by high levels of oleic, linoleic and linolenic acids.

III-10-Oxidation of lipids

For example, in hepatic or extra-hepatic reserve tissues, GAs are re-esterified into triglycerides for storage in adipocytes. In muscles of all types, they are mainly oxidized to produce energy in the form of ATP.

After digestion, absorption, transport and/or biosynthesis of these hydrocarbon molecules, the fate of these fatty acids varies greatly depending on the tissues in which they are found.

Fish cells are doubtful for peroxisomal β-oxidation in hepatocytes.

Extensive studies in cod and haddock show that peroxisomal β-oxidation accounts for 30% of total hepatic fatty acid oxidation (Crockett and Sidell, 1993; Nanton et al., 2003). But it is more limited in white muscle, where mitochondrial β-oxidation generally predominates (Nanton et al., 2003).

IV-Benefits of PUFA consumption

IV-1-Effects of PUFAs on the cardiovascular system

Consumption of fish oil or EPA and DHA supplements has favourable or neutral overall effects on the main cardiovascular risk factors. Oral administration of these PUFAs produces spectacular effects on the circulatory system (lower plasma triglycerides, higher HDL-cholesterol, stability of total cholesterol, improved vasodilation or vasoconstriction, lower systolic and diastolic blood pressure). These effects are highly dependent on the quantity of fatty acids ingested (Geleijnse et al., 2002; Dickinson et al., 2006; Balk et al., 2006).

Other studies have shown that fish oil and the n-3 LC-PUFAs EPA and DHA have a proven impact on activity and heart rate (Mozaffarian et al., 2006).

These results are consistent enough across studies to suggest that these biomolecules reduce the fatal complications of myocardial infarction (Mozaffarian and Rimm, 2006).

However, n-3 LC PUFAs offer no benefit in preventing relapses of heart failure (Tavazzi et al., 2008).

All in all, a daily intake of 500 mg of EPA and DHA seems justified for the general population, taking all age groups into account. This is with a view to preventing most cardiovascular problems. This hypothesis is supported by the fact that atherosclerotic lesions can also appear during childhood and even adolescence (Berenson et al., 1998).

This has also been demonstrated in recent studies (Chavarro et al,. 2007; Hedelin et al,. 2007), but there is still some heterogeneity, although the number of studies showing a reduction in this risk is increasing.

Only the Japanese study by (Hirose et al., 2003) showed a reduction in risk with high fish consumption. Nevertheless, the effects of omega-3 and omega-6 are not restricted to the cardiovascular system. In fact, they help prevent and correct a number of physiological and psychological disturbances.

IV-2-Hepatic steatosis

Preliminary work has shown that consumption of n-3 LC-PUFAs improves the hepatic steatosis that accompanies metabolic syndromes (Capanni et al., 2006; Zivkovic et al., 2007).

IV-3-Age-related macular degeneration

A Japanese analysis of 9 epidemiological studies shows that a high intake of LC-PUFAs is associated with a 30% reduction in the risk of late-onset age-related macular degeneration (AMD). Regular and continuous consumption of fish at least 2 times a week has been shown to be associated with a reduced risk of both early and late AMD (Chong et al., 2008).

IV-4-AGPI and depression

In psychiatric settings, a drop in the percentage of n-3 PUFAs, particularly long-chain n-3 PUFAs (LC n-3 PUFAs), has often been observed in plasma or erythrocyte lipids in depressed patients, when compared to control subjects. This is also demonstrated by the positive correlation between increased plasma levels of omegas 3 and 6 and reduced depression (Adams et al., 1996 ; Maes et al., 1996 ; Peet et al., 1998 ; Frasure-Smith et al., 2004).

However, it is known that the CNS and retina are enormously rich in lnc-PUFAs: EPA, ARA and DHA. It is therefore dependent on these molecules for its proper functioning and development.

Asian (Japan, Korea, Taiwan) and Mediterranean countries have both very high fish consumption and a low prevalence of depression (Hibbeln, 1998). Depression is a major public concern, as it increases mortality (Cuijpers and Smith, 2002) and is a risk factor for chronic diseases such as cardiovascular disease (Van der Kooy et al., 2007) and Alzheimer's disease (Ownby et al., 2006).

IV-5-AGPI and schizophrenia

Clinical intervention trials with n-3 PUFA-LC have been carried out in schizophrenic subjects. Uncontrolled trials of these compounds in isolated patients, also untreated by other antioxidants, have shown significant and rapid improvements (within two months) (Puri et al., 2000).

IV-6-Stress and aggression

A few controlled trials have been carried out to test the effect of n-3 PUFA supplementation on aggression in stressed and unstressed subjects. Daily supplementation for 2-3 months with 1.5 g DHA + 0.2 g EPA prevented aggressive tendencies in adults under natural or experimental stress (Hamazaki et al., 1998; Hamazaki et al 2005).

IV-7-Asthma and bronchoconstriction

A number of arguments point to the role of omega-3s in the prevention of asthma and airway allergy (Stephensen, 2004; Mickleborough and Rundell, 2005). At high doses, n-3 LC PUFAs could play a spectacular role in reducing exercise-induced bronchoconstriction.

IV-8-Adult, mother and newborn

Several prospective observational studies suggest that inadequate intakes of long-chain PUFAs, particularly n-3 PUFAs, during pregnancy and lactation, may be associated with increased risk later in life. Psychopathological disorders in children have been observed (autism, ADHD, social behavioural defects), which have no nutritional causal link. It is well known that large quantities of n-6 and n-3 PUFAs are necessary for the development of the newborn, as they play an active role in the construction of cell membranes and the development of the brain and central nervous system. This mainly concerns arachidonic acid (AA, 20:4 n-6) and DHA.

In vivo, since the biosynthetic activity of these GAs is too low to meet requirements, it is necessary to supplement the newborn's diet with these GAs. To this end, one approach has been to use fish oils in milks. With this type of supplementation, DHA levels in plasma and in the membrane phospholipids of red blood cells are close to those observed with breast milk (Carlson et al., 1992). However, EPA deficiency causes a reduction in endogenous AA synthesis and impairment of many physiological reactions (Neddelman et al., 1979; Lee et al., 1985). For adults, Table 2 shows daily GA recommendations, which can be expressed as follows:

- SFA: 25% of total lipid intake.

- MUFA: 60% of total lipid intake.

- PUFA: 15% of total lipid intake, with an optimal n-6 / n-3 ratio of 5.

All in all, a supplement or diet rich in n-3 PUFAs, particularly EPA, tends to attenuate physiological responses, correct biological disturbances and could modulate cognitive and psychological responses to stress in healthy volunteers.

Table 2: Recommended Dietary Allowances (RDA) for fatty acids (Martin, 2001)

		Homme adulte		Femme adulte	
Apport énergétique total [AET] (kcal / j)		2 200		1 800	
		g / j	% AET	g / j	% AET
Acides gras saturés (AGS)		19,5	8	16	8
Acides gras monoinsaturés (AGMI)		49	20	40	20
A	Ac. Linoléique (C18 : 2 n-6)	10	4	8	4
G	Ac. Linolénique (C18 : 3 n-3)	2	0,8	1,6	0,8
P	AGPI-LC	0,5	0,2	0,4	0,2
I	dont DHA	0,12	0,05	0,1	0,05
	Apport lipidique total	81	33	66	33

AGPI : Acides Gras PolyInsaturés   *AGPI-LC : AGPI à Longue Chaîne*   *DHA : Ac. DocosaHexaenoïque (C22 : 6 N-3)*

Materials and methods

I-Sampling

I-1-Biological material

The present work was based on 3 batches of a single labrid species, Symphodus *tinca* (Linnaeus, 1758), the most common in Tunisian waters (Baradai, 2000). Each batch comprises 12 specimens, with 6 males and 6 females examined at each site.

Our biological material was collected during spring (March), which is included in the species' reproductive period, and from 3 different stations:

- Marine station: Northwest (Tabarka).

- Island station: Southeast (Djerba)

- Lagoon station: North-east (Ghar Elmelh lagoon).

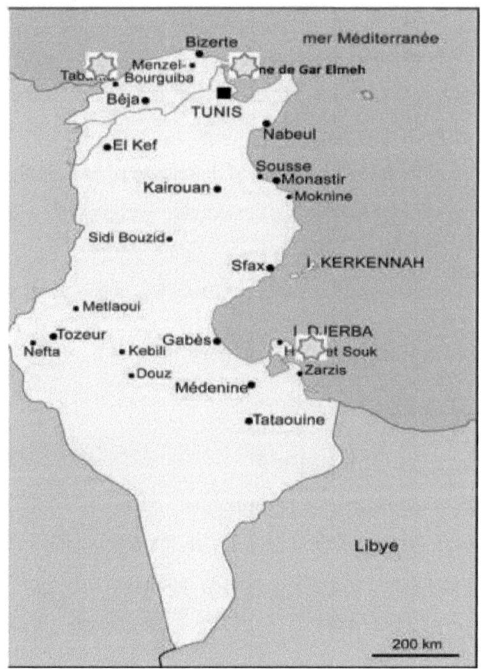

Source(Google.com.).

I-2-Precautions and storage

Sampling was carried out with care, choosing fish in good condition, not crushed, and above all fresh to obtain good results, whether for morphometric measurements or the quality and quantity of lipids in the various organs. Once captured, the fish are quickly brought to the laboratory at low temperatures, in a cool box or isothermal bag to avoid any biodegradation of the tissues.

Although there are at least two types of fish spoilage: bacterial and enzymatic (Uchyama and Ehira, 1974).

By lowering the temperature to around 0°C, the development of spoilage and pathogenic micro-organisms is reduced. This reduces the rate of spoilage of fish by these microorganisms. In the past, lowering the preservation temperature also reduced the speed of enzymatic reactions.

In the laboratory, fish are processed immediately or kept refrigerated at -40°C until analysis to avoid any alteration in lipid composition.

In addition, during the analysis and dissection of each individual, fish parameters were determined: total weight (Pt), eviscerated weight (Pev), liver weight (Pf) and gonad weight (Pg).

All measurements were carried out using a precision balance for weight parameters.

II-Lipid composition

II-1-Total lipid extraction

1-1-Principle of the method

Extraction is based on the hydrophobicity of lipids and their miscibility in chloroform. The principle of the method is based on the combination of a polar solvent, methanol, which destabilizes hydrogen bonds, and a non-polar solvent, chloroform, which entrains membrane lipids.

1-2-Extraction

Total lipids are estimated using the method of Floch et al (1957), modified by Bleyh and Dyer (1959). Fillets of liver and gonad flesh weighing 1 g are extracted with chloroform-methanol (2:1 V/V), 0.01% BHT and 15% NaCl solution. In fact, each specimen is completely ground in a porcelain mortar with the gradual addition of 30 ml of the chloroform solution. Then, 8 ml of NaCl solution is introduced to help separate the reject protein phase. The total mixture is then centrifuged at 4000 rpm for 20 minutes. Finally, 3 phases are obtained:

- An upper aqueous phase containing salt water, proteins and lipoproteins.

- An intermediate tissue phase.

- A chloroformed lower phase containing total lipids.

The protein phase is retained with a Pasteur pipette and discarded, while the chloroform phase is recovered with a syringe and transferred to a flask of known empty weight PV. This extract is evaporated completely using a steam rotator. The temperature is kept at 40°C to recover the dry adipose deposit.

The weight of oily residue is therefore the difference between the empty weight of the flask (PV) and its final dry weight (Pf). Thus, the lipid content of 100 grams of fresh sample material is estimated according to the following formula

$$\textbf{Lipids (\%) = pf - PV /p (organ)} * \textbf{100}$$

With :

-Pi: initial weight of the balloon.

-Pf: final weight of the flask containing the dried oil fraction.

-P (organ): weight of the organ to be analyzed (flesh, liver or gonad).

To avoid any deterioration of the dried fats, 2ml of chloroform or toluene-ethanol mixture (4:1, V:v) is added to each pellet, then the extracts are placed in a refrigerator at -40°C.

II-Methylation

Or trans-esterification reaction, which cuts the bonds between the glycerol molecule and the three fatty acids for TAGs, and between the fatty acid and the fatty alcohol for waxes. This reaction releases fatty acids that are methylated and re-esterified by methanol.

II-1-Obtaining fatty acid methyl esters

The method adopted to obtain methyl esters is that of Cecchi et al. (1985). Its principle is the trans-esterification of fatty acids.

II-2-Experimental protocol

-Take 20 µL of the preserved lipid extract and dry with nitrogen gas.

- Add 2 ml hexane to ensure solubility of GAs in the organic phase.

Add 0.5 ml sodium methylate to make the GAs volatile in the GC column.

- Vortex the mixture and leave for 2 minutes to dilute the GAs in the organic phase.

Adding 0.2 ml H_2 SO_4 followed by 1.5 ml NaCl (15%) enables separation of the aqueous reject phase.

-Centrifuge at 2000 rpm for 10 minutes and recover the supernatant.

-After drying with nitrogen, between 100 and 300 µL of hexane are added, depending on the concentration of the mixture.

III-Fatty acid separation

After methylation, the GAs are analyzed by gas chromatography using a brand-name chromatograph (Agiles Technologies, Santa Clara, CA, USA) model 6890 N equipped with a F.I.D (flame ionization detector) and an INNOWAX capillary column; dimensions: 30 m long and 0.25 mm internal diameter, film thickness 0.25 mm (Agiles Technologies). The injector is an Intel split or splitless capillary. Detector temperature is set at 280°C, while injector temperature is maintained at 230°C.

This chromatography separates the fatty acid molecules contained in the methylated extract, rendering the various compounds volatile by heating without decomposing them.

The 1 ul volume extract is placed at the head of the column via a micro-syringe, migrating to the injector which is a chamber located at the top of the column. The flow rate of nitrogen as carrier gas was set at 1.5 ml/min in an oven at a constant temperature of around 150°C for 1 min, then raised gradually by programming at a rate of 15°C/min to 210°C. The temperature was then held constant for 5 min and finally programmed to 250°C at a rate of 4°C/min. This temperature increase served to evaporate the extract.

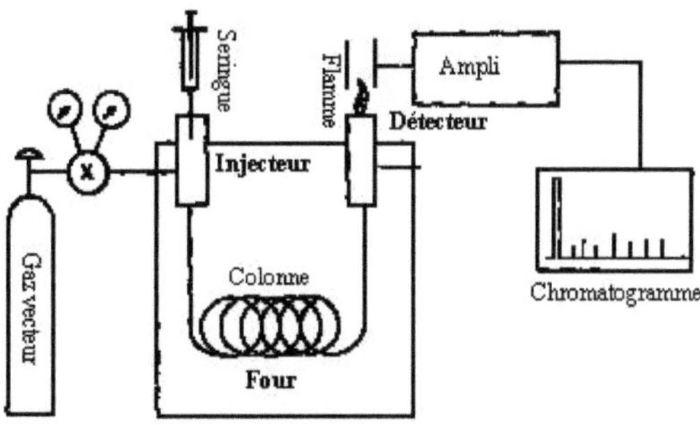

Figure 10: Components of a chromatograph (FID)

Thus volatilized, the various molecules are transported by this gas to the stationary phase, where they are retained. At this point, all the molecules are separated according to their affinity with this phase (Figure 10).

The constituents are all retained and released separately: so the compound with the least affinity for the stationary phase would spend a minimum amount of time leaving the column.

On leaving the column, the molecules encounter the F.I.D. detector, which is highly sensitive to organic compounds. Next, the effluent is burnt by an air-hydrogen flame. Combustion releases carbon ions, which are collected by electrodes surrounding the flame.

The electrical charge of the ions is amplified by an electrometer, transforming them into a voltage which is sent to a recorder, informing it of the quality of each component and its various physical properties.

Analyses were performed on a Hitachi 912 analyzer (Roche Diagnostics GmbH, Mannheim, Germany). A chromatogram is generated for each molecule, consisting of several curves whose peaks vary according to the intensity of the molecule being analyzed.

IV-Identification of fatty acids

The analysis of the chromatograms obtained is relatively complex, given the quantity of fatty acids present in the samples. In fact, each sample analyzed contains an average of fifteen different fatty acids.

To determine retention times and factors specific to each ester, we injected several standard solutions and calculated the averages, which we then fed into the integrator or recorder.

Thus, to identify with certainty each of the fatty acids present, we compared the retention times of the sample compounds with those of the PUFA No. standard mixture.

To be able to quantify fatty acids in the samples to be analyzed, we first need to draw up calibration curves for the various compounds. What's more, before using these curves, we need to check that they are linear.

At the output, the chromatogram will provide a series of peaks of varying separation, size and width. Depending on the detection method, the area of a peak is proportional to the quantity of product represented by that peak. By measuring the area of each peak and relating it to the total area of all the peaks, we can determine the percentage of each of the components contained in the analyzed mixture.

Retention times and peak areas are determined from the recorded file using a dedicated computer program. It prints out an analysis report giving retention times and areas of each peak. This is a small computer that retrieves all the data from the detectors, plots the chromatograms and integrates the peak areas. It can be programmed to perform the various calculations leading to concentrations from standard chromatograms and chromatograms of the mixtures analyzed.

When the peaks are poorly separated, the area of the individual peaks can be approximated, and concentrations can still be estimated.

Results

I-Comparisons of hepatho-somatic and gonado-somatic ratios between males and females of *Symphodus tinca* from the three stations (month of March)

Table 3 shows the variations in RGS and RHS of male and female *Symphodus tinca* in the three environments (marine, island and lagoon), during the month of March.

Table 3: Comparison of hepatho-somatic and gonado-somatic ratios between males and females of *Symphodus tinca* from the three stations.

	Females marine	Males sailors	Females islanders	Males islanders	Females lagoon	Males lagoons
RGS	3,90 ± 0,27	1,54 ± 0,19	3,97 ± 0,41	1,61 ± 0,02	2,5 4 ± 0.72	1,44 ± 0,11
RHS	0,89 ± 0,03	1,64 ± 0,34	0,67 ± 0,08	1,23 ± 0,03	0,69 ± 0,05	1,27 ± 0,18

For all environments (marine, lagoon and island), the RGS of the females studied is higher ($♀$I= 3.97 ± 0.41; $♀$M = 3.90 ± 0.27 and $♀$L= 2.54 ± 0.72) than that of the males ($♂$I= 1.61 ± 0.02; $♂$M= 1.54 ± 0.19and $♂$L=1.44 ± 0.11). Also, we note that island individuals indicate the highest RGS, the average for marine individuals and the lowest for lagoon specimens.

the results showed that whatever the population studied, males had the highest (RHS) compared with females ($♂$I=1.23 ± 0.03; $♀$I= 0.67 ± 0.08) and ($♂$M=1.64 ± 0.34; $♀$M= 0.89 ± 0.03). ($♂$L= 1.27 ± 0.18; $♀$L=0.69 ± 0.05). In the insular polulation, we note that they are specified by the lowest RHS ($♀$I= 0.67 ± 0.08; $♂$I =1.23 ± 0.03) and the highest RGS of all the batches examined ($♀$I= 3.97 ± 0.41; $♂$I= 1.61 ± 0.02), (Table 3).

II-Quantitative study of tissue lipids

II-1- Comparison of the total lipid content of *Symphodus tinca* from the marine population as a function of sex (month of March)

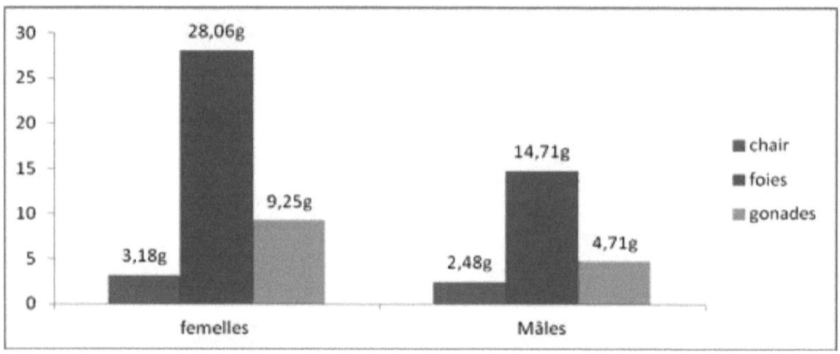

Figure 11 shows variations in lipid levels in the flesh, liver and gonads of male and female *Symphodus tinca* in the marine environment during March.

Figure 11: Total lipid levels in males and females of *Symphodus tinca* from the marine environment (month of March).

Referring to the results represented in this figure, we note that the amount of lipids contained in female muscles is estimated at: 3.18 ± 0.25g/100 g fresh matter, is higher than that represented by male fillets (2.48 ± 0.31g/100 g fresh matter).

Also, in terms of livers and gonads, female specimens show higher fat levels (28.06 ± 2.49g and 9.25 ± 0.62g/100g fresh matter) than males (14.76 ± 1.03g and 4.71 ± 0.52g/100g fresh matter), respectively for male livers and gonads. Which means that, across all organs, marine females accumulate more fat than drifting males from the same site.

II-2-Comparison of lipid profiles of *Symphodus tinca* from the island population as a function of sex (month of March)

Figure 12 shows the variations in fat content in the three organs (flesh, liver and gonads) of *Symphodus tinca* individuals (males and females) from insular coasts, during the ripening period.

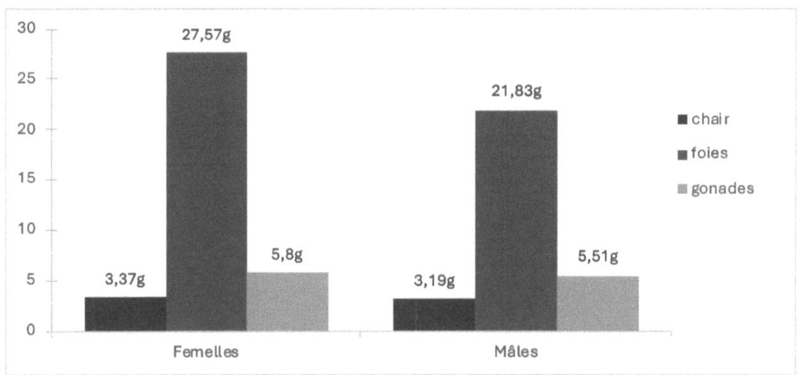

Figure 12: Total lipid levels in males and females of *Symphodus tinca* from the island environment (month of March).

Lipid levels in the flesh of males and females were similar, but higher in males. These levels are (3.19 ± 0.66g/100 g fresh matter) and (3.37 ± 0.49g/100 g fresh matter), respectively for male and female flesh.

Lipid profiles for livers and gonads show that females have higher levels than males. These levels are of the order of 27.57 ± 4.78g and 5.8 ± 0.34g/100 g fresh matter, respectively in female livers and gonads ♀. While the rates represented by males are of the order of 21.83 ± 2.35g/100 g for livers and 5.51 ± 0.62g/100 g fresh matter for gonads ♂.

II-3-Comparison of lipid profiles of male and female *Symphodus tinca* from the lagoon station (March)

Figure 13 compares the lipid content of flesh, liver and gonads between male and female *Symphodus tinca* from the lagoon environment.

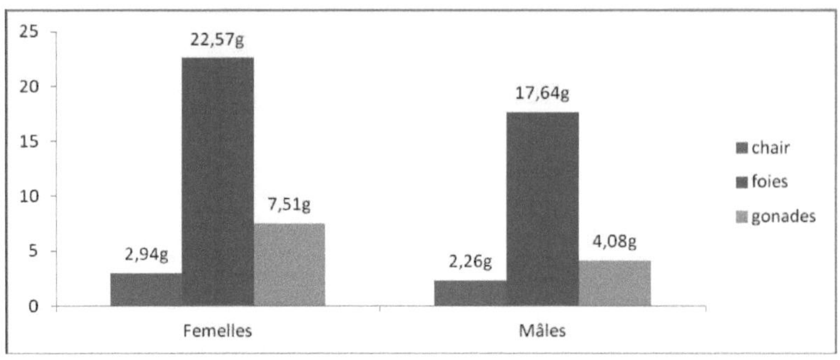

Figure 13: Total lipid levels in males and females of *Symphodus tinca* from the lagoon environment (month of March).

The established lipid profiles of the lagoon population show that peacock crenilabra females have the highest lipid levels in all three organs (Figure 13).

Flesh content: 2.94 ± 0.83 g/100 g fresh matter (females) vs. 2.26 ± 0.42 g/100 g fresh matter (males). These levels indicate that this population has the lowest muscle lipid content compared with the other 2 populations (Figures 11, 12 and 13). For livers, this fraction is around 22.57 ± 3.10g/100 g MF, for females, versus 17.64 ± 2.09g/100 g MF, for males. Also, female gonads show a rate of (7.51 ± 0.82g/100 g MF) which is higher than (4.08 ± 0.90g/ 100 g MF) in male gonads.

For both sexes, livers contain more lipids than gonads and tenderloins.

This quantitative arrangement of lipid contents is similar to that of the lipid profile of the marine and island population, with liver and gonads representing the highest levels compared to fillets (Figures: 11,12 and 13).

II-4-Comparison of lipid profiles of *Symphodus tinca* females from the three environments (month of March)

Figure 14 compares the lipid content of flesh, liver and gonads in maturing *Symphodus tinca* females from three populations: marine, lagoon and island.

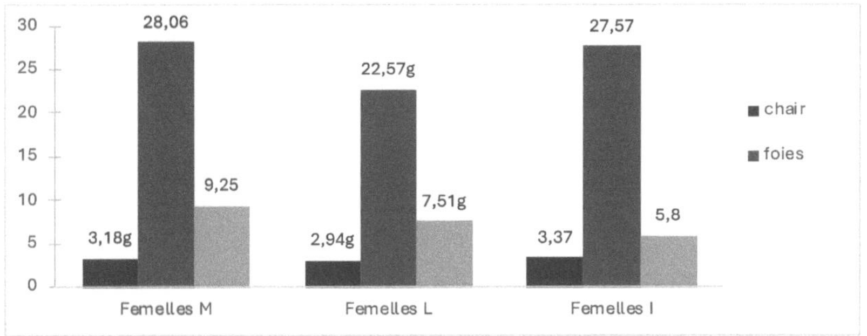

Figure 14: Toal lipid levels in *Symphodus tinca* females from marine, lagoon and island environments (month of March).

Based on the values in this figure, it can be seen that the flesh of island and marine females has similar lipid levels, at 3.18 ± 0.25g and 3.37 ± 0.49g/100 g MF respectively. These levels are higher than those of the lagoon environment, where the content is 2.94 g±/100 g MF.

Livers from marine and island females show similar levels, at 28.06 ± 2.49 g and 27.57 ± 4.78g/100 g MF respectively. On the other hand, the minimum rate is 22.57 ±3.10g/100 g MF, recorded for the livers of lagoon individuals.

Thus, lipid mobilization is more advanced in marine and island females, arguing in favor of earlier maturation than in lagoon-derived females.

During maturation, the gonads of female specimens from the marine environment have the highest lipid content (9.25 ± 0.62g/100 g FFM), while those of females from the lagoon population have the lowest (7.51 ± 0.82g/100 g FFM). Ovaries from island females, on the other hand, had the lowest lipid content (5.80 ± 0.34g/100 g MF).

II-5-Comparison of the lipid profile between Symphodus tinca males from the three environments (month of March)

Figure 15 shows a comparative study of the lipid profiles of the livers and gonads of three batches of males from the marine, lagoon and island populations in the maturation period (month of March).

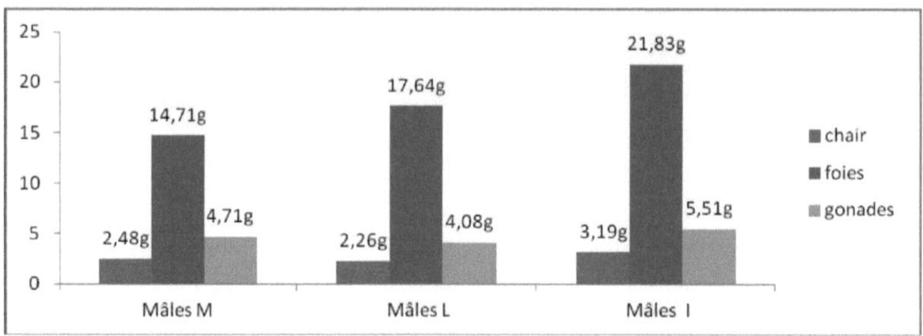

Figure 15: Total lipid levels in *Symphodus tinca* males from marine lagoons and islands (March).

Lipid content in the flesh of males from the three environments shows the following variations: The highest content is recorded for island individuals, at around 3.19 ±0.66g/100g MF, the average for marine specimens (2.48 ± 0.31g/100g MF) and the lowest for lagoon males (2.26 ± 0.42g/100g MF).

Comparison of the 3 lipid profiles of male livers shows that island males have the highest content compared with the others (lagoon and marine). The contents are: 21.83 ± 2.35g; 17.64 ± 2.09g and 14.71 ±1.03g /100g MF, approximately for island, lagoon and marine males. During the maturation period, male gonads from the lagoon population have the lowest lipid content (4.08 ± 0.90g/100g MF), compared with the other two, from the island station (5.51 ± 0.62g) and the marine station (4.71 ± 0.52g/100g MF).

III-Qualitative analysis of fatty acids

Gas chromatography analysis of methylated extracts from the various organs identified 15 long-chain fatty acids present in the total lipids of *Symphodus tinca* flesh, liver and gonads (Appendix 2).

III-1-Comparison of fatty acid composition between males and females of the *Symphodus tinca* marine population (month of March)

1-1-Comparison of flesh fatty acid composition between individuals (♂+♀) of the marine population

According to Figure 16, the flesh of all individuals in the northern marine environment shows the same quantitative arrangement of fatty acid classes:

Polyunsaturated fatty acids are the dominant TFA class, with proportions of 55.58% for females and 47.83% for males.

So, whether male or female, muscle lipids are composed primarily of PUFAs, secondarily of SFAs and to a lesser extent of MUFAs.

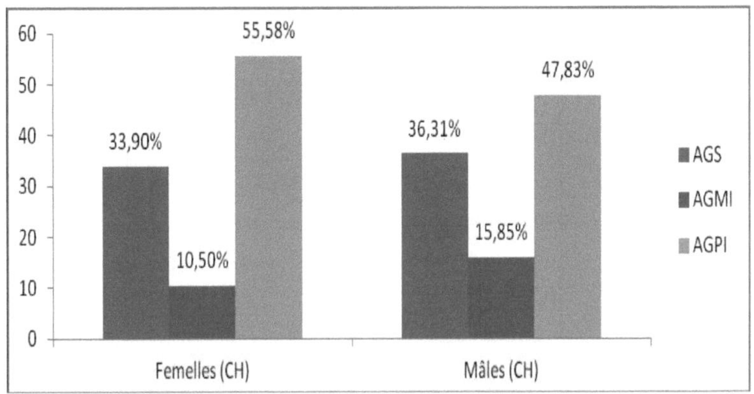

Figure 16: Variations in percentages of saturated (SFA), monounsaturated (MUFA) and polyunsaturated (PUFA) fatty acids in the flesh of the *Symphodus tinca* marine population as a function of sex (month of March).

Flesh lipids of female individuals from the Tabarka coast show the lowest percentages of SFA when compared with those of males from the same station. These rates are: 36.31% and 33.90% for males and females respectively (Figure 16). However, the flesh of female specimens shows the highest PUFA content compared with that of males, at 55.58% and 47.83% for females and males respectively.

With regard to MUFAs, the rate represented by the males' TFAs is 15.85%, higher than that of the females' fillets: 10.50%.

1-2-Comparison of the fatty acid composition of the livers of individuals (♂+♀) from the marine population.

In contrast to the lipid profiles of the flesh, the majority class of liver FAs is SFAs, with a smaller proportion of PUFAs and MUFAs (Figure 17). Nevertheless, there is a gender-dependent variation in liver FA composition: female marine specimens have the highest percentages of SFAs (46.01%) and PUFAs (41.33%), compared with males from the same station, whose SFA and PUFA contents are 40.28% and 35.16% respectively.

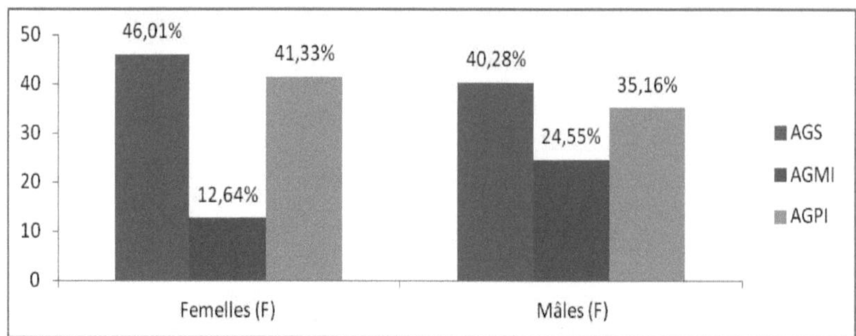

Figure 17: Variations in percentages of saturated (SFA), monounsaturated (MUFA) and polyunsaturated (PUFA) fatty acids in the livers of the *Symphodus tinca* marine population as a function of sex (month of March).

The two hepatic fat profiles show a significant variation in MUFA levels: hepatic lipids in females are very low, at around 12.64%, compared with around 24.5% in males.

1-3-Comparison of gonadal fatty acid composition of individuals (♂+♀) from the marine population.

Comparative study of the GA composition of gonads from marine specimens shows that, as with their flesh, PUFAs also present the highest level in total lipids extracted from male and female gonads. These levels are estimated at 58.25% for ♀ gonads, and 51.94% for ♂ gonads (Figure 18).

TFAs in male gonads are higher than those in ovaries, while SFA levels are similar and slightly higher in males. IMFA and SFA levels are : 10.79% and 31.95% for female gonads; 15.30% and 32.74% for male gonads (Figure 18).

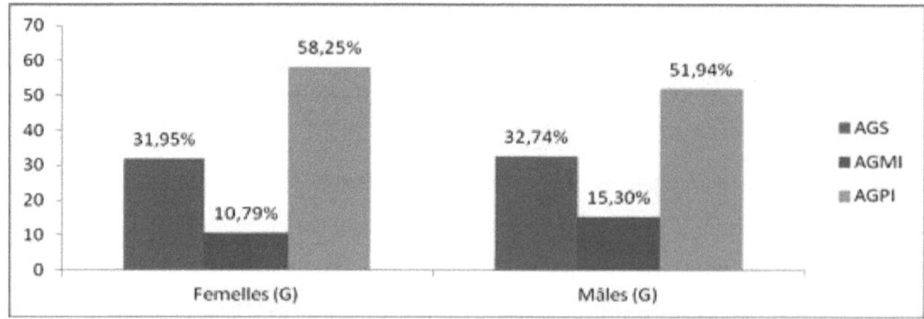

Figure 18: Variations in percentages of saturated (SFA), monounsaturated (MUFA) and polyunsaturated (PUFA) fatty acids in the gonads of the *Symphodus tinca* marine

Table 4 shows the variations in PUFA percentages for the three organs of the marine population. It can be seen that the organ richest in PUFAs is the gonads, followed at $2^{\text{ème}}$ level by the flesh and, to a lesser extent, the liver. Female subjects also report higher PUFA levels in 3 organs (Table 4).

Table 4: Variations in percentages of PUFAs in the flesh, livers and gonads of *Symphodus tinca* living in marine waters as a function of sex (moi de Mars).

% TFA	Females	Males
Chair	55,58%	47,83%
Liver	38,60%	30,38%
Gonads	58,25%	51,94%

1-4-Comparison of major fatty acids of individuals (♂+♀) from the marine population.

Static analysis of flesh, liver and gonad TFAs allows us to express the corresponding majority fatty acid for each fatty acid family.

Table 5 shows the variations in major fatty acid percentages for fillets from the *Symphodus tinca* marine population as a function of sex.

Table 5: Variations in the percentage of major fatty acids in fillets from the *Symphodus tinca* marine population as a function of sex (month of March).

% TFA	Females	Males
Major SFA (palmitic acid)	23,17%	26,81%
Major MUFA (oleic acid)	7,58%	10,99%
Major PUFA (DHA)	23,30%	15,66%

For TFAs in the flesh and gonads of marine individuals, the majority of each fatty acid class is C16:0, C18:1 n-9 and C226 n-3 (Table 5).

In the flesh of males, C16:0 is the most abundant TFA. Its 26.18% content is higher than that of DHA and oleic acid, which have DHA and C18n-9 contents of 15.66% and 10.99% respectively.

In contrast to the lipids in male fillets, DHA and C16:0 levels are similar in female flesh. These levels are higher than that of oleic acid, with percentages in the order of : 23.30%, 23.17% and 7.58% for DHA, C16:0 and C18:1n-9 respectively, making DHA the most abundant fatty acid in female flesh (Table 5).

Table 6 shows the variations in major fatty acid percentages for liver TFAs in the *Symphodus tinca* marine population as a function of sex. For both sexes, palmitic acid is the major fatty acid of the TFAs, with a higher level in females (29.00%) than in males (26.37%).

Table 6: Variations in the percentage of major fatty acids in the livers of the *Symphodus tinca* marine population as a function of sex (month of March).

% TFA	Females	Males
AGS major (C16:0)	29,00%	26,37%
Major MUFA (C18:0)	7,96%	14,75%
Major PUFAs	(DHA)17.30% OF	(EPA)11.18

Oleic acid, on the other hand, has a higher level in the livers of females (14.75%) than in those of males (7.96%) (Table 6). The major PUFA also varies according to sex: it is represented by DHA in the liver lipids of females and by EPA in those of males.

Table 7 shows the variations in major fatty acid percentages for gonadal TFAs in the *Symphodus tinca* marine population as a function of sex.

The major fatty acids in the gonads of marine specimens are: C16:0, C18:1n-9 and C22:6n-3. This result is similar to that of MFA from fillets (Table 5, 7).

Table 7: Variations in gonadal major fatty acid percentages of the *Symphodus tinca* marine population as a function of sex (month of March).

% TFA	Females	Males
AGS major (C16:0)	24,99%	24,78%
Major MUFA (C18:n9)	8,89%	9 ,15%
Major PUFA (DHA)	18,20%	15,05%

As for liver TFAs, palmitic acid accounts for the majority of all FFAs. It has similar percentages in male and female gonads. These are 24.99% and 24.78% for female and male livers respectively (Table 7).

This quantitative distribution is similar for oleic acid, with levels of 8.89% in female and 9.15% in male gonads (Table 7). DHA, on the other hand, is estimated at 18.20% in the ovaries, higher than in the male gonads (15.05%).

III-2-Comparison of fatty acid composition of males and females from the *Symphodus tinca* lagoon population (month of March)

2-1-Comparison of flesh fatty acid composition between individuals ($\male+\female$) of the lagoon population.

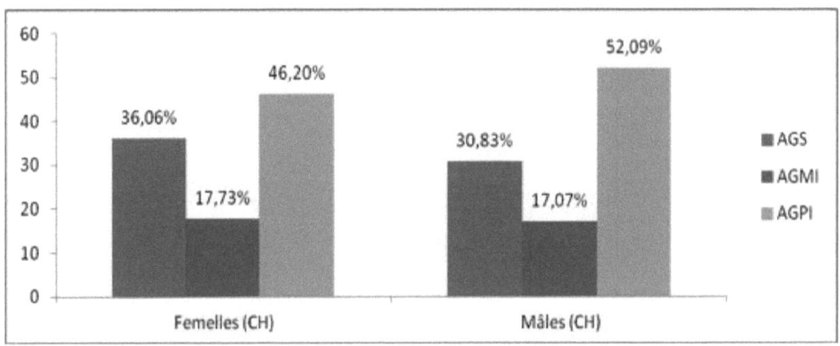

Figure 19:Variations in percentages of saturated (SFA), monounsaturated (MUFA) and polyunsaturated (PUFA) fatty acids in the flesh of the *Symphodus tinca* lagoon population as a function of sex (month of March).

According to Figure 19, the fatty acid profile of the flesh of lagoon specimens shows a composition similar to that of individuals from the marine environment (Figure 16). In fact, the most dominant fraction of total muscle lipids is GLA, with medium proportions of SFA and low proportions of MUFA.

Table 8 shows a comparison of (PUFA) percentages in the flesh of the two populations: marine and lagoon.

Table 8: Variations in the percentages of (PUFAs) in the flesh of both marine and lagoon populations of *Symphodus tinca* as a function of sex (month of March).

% TFA	Females marine	Females lagoons	Males sailors	Males lagoon

PUFA content nets	55,58%	46,20%	47,83%	52,09%

The TFAs of fillets from lagoon males are richer in PUFAs than those of marine males: with 52.09% for lagoon males and 47.84% for fillets from marine males.

On the other hand, the PUFA content in the flesh of marine females is higher than in the flesh of lagoon females. The respective proportions were 55.58% (fillets from marine females) and 46.20% (fillets from lagoon females).

Lipids in the fillets of specimens from the Ghar Elmelh lagoon (Figure 19) show similar levels of MUFAs, but slightly higher levels of TFAs in males (17.73%) than in females (17.07%).

On the other hand, female fillets have higher SFA percentages than males. These are 36.06% and 30.83% respectively.

2-2-Comparison of the fatty acid composition of the livers of individuals (♂+♀) from the lagoon population.

Figure 20 shows the variation in fatty acid percentages in liver tissue for specimens from the Ghar Elmelh lagoon. TFAs in male livers are richer in SFAs, but with similar levels of MUFAs, when compared with liver TFAs in females.

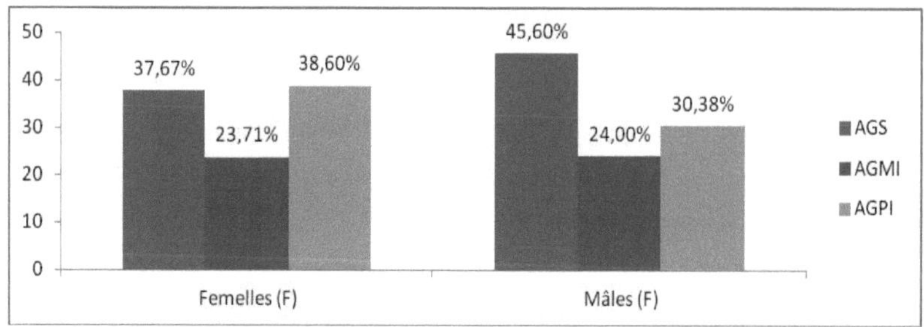

Figure 20: Variations in percentages of saturated (SFA), monounsaturated (MUFA) and polyunsaturated (PUFA) fatty acids in the livers of the *Symphodus tinca* lagoon population as a function of sex (month of March).

The contents are : 45.60% and 37.67%, for SFAs, 24.0% and 23.71%, for MUFAs; respectively, for male and female livers.

Variations in hepatic PUFA levels in the marine and lagoon populations are shown in Table 9.

Table 9: Variations in polyunsaturated fatty acid (PUFA) percentages in the livers of two populations: marine and lagoon *Symphodus tinca*, as a function of sex (month of March).

% TFA	Females marine	Females lagoons	Males sailors	Males lagoons
Hepatic PUFA levels	41,33%	38,60%	35,16%	30,38%

The TFA content of female livers is higher than that of males. We also note that the livers of lagoon-dwelling females accumulate more PUFAs than those of marine females. Also, hepatic PUFAs show notable variations for both populations. These variations are elucidated in Table 10.

Table 10: Comparison of the percentages of *Symphodus tinca* livers in the marine and lagoon populations according to sex (month of March).

% TFA	Females marine	Females lagoons	Marine males	Males lagoons
MUFA content of livers	%12,64	%23,71	%24,55	%24,00

The liver lipids of males contain more MUFA than those of females: the percentages are estimated at 12.64% for female livers and 24.55% for male livers. These results are different for the marine population, where IMFA levels in livers are similar for both sexes (Table 10).

2-3-Comparison of gonadal fatty acid composition of individuals (♂+♀) from the lagoon population.

Figure 21 shows the variations in the content of each class of FA in the male and female gonads of lagoon specimens. From these percentages, we conclude that: gonadal TFAs of lagoon individuals testify to a dominance of polyunsaturated fatty acids (PUFA:♀42.68%; ♂59.30%), over saturated (SFA:♀33.80%; ♂27.55%) and monounsaturated fatty acids (MUFA:♀23.51%; ♂13.14%). These results are similar to flesh and gonad lipids in the marine population.

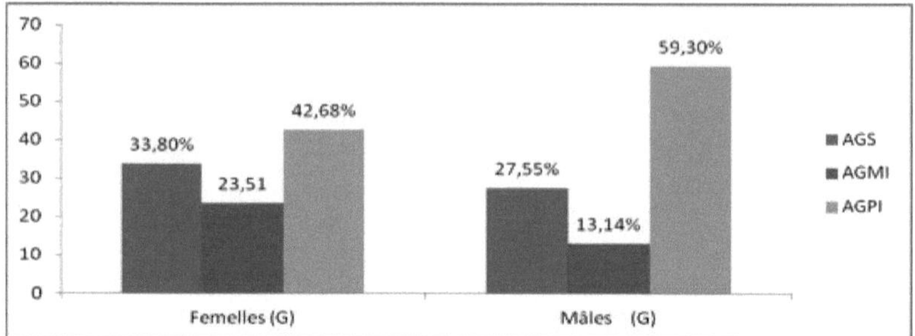

Figure 21: Variations in percentages of saturated (SFA), monounsaturated (MUFA) and polyunsaturated (PUFA) fatty acids in the gonads of the *Symphodus tinca* lagoon population as a function of sex (month of March).

TFAs in male gonads are the richest in PUFAs and determine the lowest level of SFAs compared to female gonads. Their contents are respectively: 59.30% and 42.68% for PUFAs; 27.55% and 33.80% for SFAs. In addition, the level of MUFA represented by ♀ gonads (23.51%) is higher than that of ♂ gonads (13.14%).

We also note that treated males have the highest levels of PUFAs in their flesh and gonads. However, this is not the case for the marine population, since females are characterized by flesh and gonads richest in PUFAs (Table 11).

Table 11: Comparison of flesh and gonad (PUFA) percentages for *Symphodus tinca* marine and lagoon populations, by sex (month of March).

% TFA	Females marine	Males sailors	Females lagoons	Males lagoons
Chair	55,58%	47,83%	46,20%	52,09%
Gonads	58,25%	51,94%	42,68%	59,30%

2-4-Comparison of major fatty acids of individuals (♂+♀) from the lagoon population.

Table 12 shows the percentages of major fatty acids in fillets from the lagoon population.

Table 12: Variations in percentages of major fatty acids in fillets from the *Symphodus tinca* lagoon population according to sex (month of March).

% TFA in flesh	Females	Males
Major SFA (palmitic acid)	23,67%	21,81%
Major MUFA (oleic acid)	12,50%	10,17%
Major PUFA (DHA)	17,67%	26,83%

In the flesh of lagoon females, palmitic acid content is around 23.67%, higher than that of DHA (17.67%). Palmitic acid is the most predominant of the TFAs. Oleic acid levels are low, at 10.17% (males) and 12.50% (females) (Table 12).

This level (in C16:0) is higher than that found in the flesh of males (21.81%). The most predominant TFA in male flesh is DHA, with a very high level compared to female flesh, at around 26.83%.

Variations in the percentages of major fatty acids in the livers of the lagoon population are shown in Table 13. Palmitic acid is the major fatty acid in the liver TFAs of all individuals, with a higher level in males (30.69%) than in females (24.35%).

Table 13: Variations in the percentages of major fatty acids in the livers of the *Symphodus tinca* lagoon population as a function of sex (month of March).

% TFA	Females	Males
Major SFA (palmitic acid)	24,35%	30,69%
Major MUFA (C16:1n7)	12,73%	13,30%
Major PUFA (EPA)	13,12%	7,03%

Palmitic acid was higher in the livers of males (30.69%) and females (24.35%) (Table 13). The major MUFA of hepatic TFAs is C16:1n7, which is higher in male livers; the results showed that the major PUFA of livers is EPA, in females with a TFA content of 13.12%, which is higher than in males (7.03%). Variations in the percentages of each major fatty acid in the gonads of the lagoon population are shown in Table 14.

Table 14: Variations in gonadal major fatty acid percentages of the *Symphodus tinca* lagoon population as a function of sex (month of March).

%in AGT	Females	Males
Major SFA (palmitic acid)	23,70%	21,93%
Major MUFA (C18:1 n9)	13,45%	8,30%
Major PUFA (DHA)	13,71%	21,38%

As with the gonads of marine specimens, the majority fatty acids in the gonadal lipids of the lagoon population are C16:0, C18:1n-9 and C22:6n-3 (Table 11 and 14).

Palmitic acid is the majority of all GAs, and is present in higher percentages in female than in male gonads. These levels are 23.70% and 21.93% for females and males respectively.

Similarly, oleic acid percentages are higher in female gonads than in male gonads, estimated at 13.45% for female gonads and 8.3% for male gonad TFAs.

On the other hand, DHA levels in male gonads are around 21.38%, higher than those in female gonads, estimated at 13.71% (Table 18). These results are similar to those for marine gonads (Tables 7 and 14).

III-3-Comparison of fatty acid composition of males and females from the *Symphodus tinca* island population (month of March)

In contrast to the fatty acid composition of the flesh of all individuals from the other two populations studied, the TFAs in the flesh of specimens derived from island coasts are made up mainly of SFAs. These are : 40.09% of TFA for female fillets and 37.00% for male fillets (Table 15).

Table 15: Comparison of (AGS) percentages of nets from three *Symphodus tinca* populations by sex (month of March).

% TFA	Marine station	Lagoon plant	Island resort
% in AGS of female flesh	33,90%	36,06%	40,09%

% in AGS of male flesh	36,31%	30,83%	37,00%

The lipid profiles of the meat from the 3 populations show that palmitic acid is the most important fatty acid in this fraction, followed by C18:0 and C14:0 (Appendix 2).

3-1-Comparison of the fatty acid composition of the flesh of individuals (♂+♀) from the island population.

Figure 22 shows the variations in the percentage of (SFAs), (MUFAs) and (PUFAs) in the flesh of the island population according to sex.

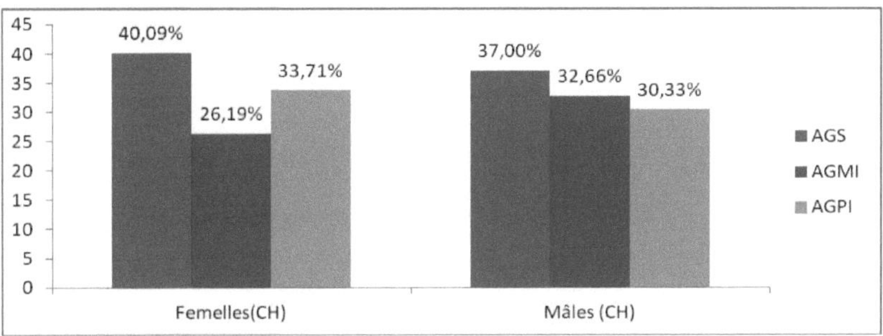

Figure 22: Variations in percentages of saturated (SFA), monounsaturated (MUFA) and polyunsaturated (PUFA) fatty acids in the flesh of the *Symphodus tinca* island population as a function of sex (month of March).

Female fillets have a PUFA content of around 33.71% of TFAs, which is higher than that of males, estimated at 30.3%. This arrangement is similar to that of the lagoon population (Figures: 19, 22).

For this population, we note that there is a quantitative variation in the disposition of each class of fatty acids in fillets that is a function of sex, as follows: TFAs in female fillets are represented primarily by SFAs, secondarily by PUFAs and to a lesser extent by MUFAs, whose percentages are 40.09%, 33.71% and 26.19% respectively.

On the other hand, the TFAs in male flesh are essentially made up of SFAs, the middle class being MUFAs and the smallest fraction being PUFAs, whose respective proportions are: 37.00%, 32.66% and 30.33% (Figure 22).

According to figure 22, the flesh of females from the island environment contains more SFAs and PUFAs than that of males. The contents are respectively: 40.09% ♀ and 37.00% ♂, in SFA; 33.71% ♀ and 30.33% ♂, in PUFA. Also, we note that the lipid composition of the flesh of island individuals marks the lowest PUFA content compared to the other two populations (Table 16).

Table 16: Variations in PUFA levels in *Symphodus tinca* fillets for the three stations (month of March).

% TFA	Marine station	Lagoon plant	Island resort
PUFA content of female flesh	55,58%	46,20%	33,71%
PUFA content of male flesh	47,83%	52,09%	30,33%

The fatty acid composition of this population shows the highest percentages of MUFA (32.6% for males and 26.19% for females) compared to the flesh of the other 2 populations (Table 17).

Table 17: Variations in MIGFA levels in *Symphodus tinca* fillets for the three species.

stations (month of March).

% TFA	Marine station	Lagoon plant	Island resort
% MUFA of female flesh	10,50%	17,73%	26,19%
% in AGMI del a male flesh	15,85%	17,07%	32,66%

3-2-Comparison of the fatty acid composition of the livers of individuals (♂+♀) from the island population.

The same is true for the FA profiles of the livers of the other two populations (Figures: 17, 20). TFAs in liver lipids from the island population are mainly represented by SFAs, and to a lesser extent by PUFAs and MUFAs (Figure 23).

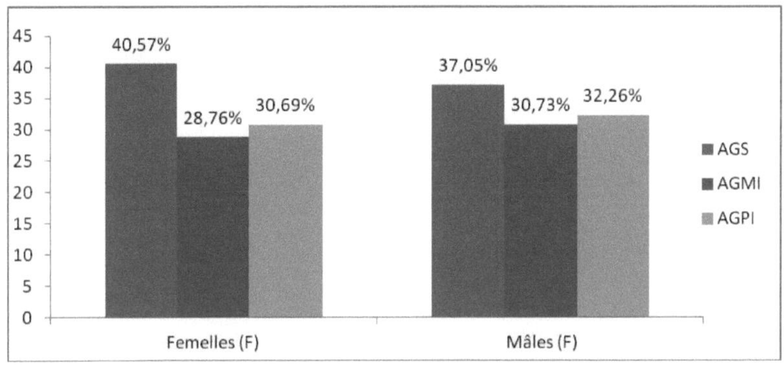

Figure 23:Variations in percentages of saturated (SFA), monounsaturated (MUFA) and polyunsaturated (PUFA) fatty acids in the livers of the *Symphodus tinca* island population as a function of sex (month of March).

Hepatic TFAs in the island population show that lipids in male livers contain more MUFAs and PUFAs than in female livers. Levels are estimated at 30.73% and 32.26% for male livers, and 28.76% and 30.69% for female livers, in IMFA and PUFA respectively (Figure 23).

In contrast, the liver lipids of female specimens show a SFA content of around 40.57%, which is higher than that of coexisting males (37.05%). This SFA arrangement is similar to that of marine livers (Figure 23).

3-3-Comparison of the fatty acid composition of the gonads of individuals (♂+♀) from the island population.

Chromatograms of gonads (♂+♀) from the island population show variations in levels for each fatty acid family. The percentages of (SFAs), (MUFAs) and (PUFAs) shown in Figure 24.

Indeed, the most dominant fraction of total lipids in these gonads is APGI, moderately so, SFAs and, in small proportion, MUFAs. This arrangement is similar to that of the other two populations.

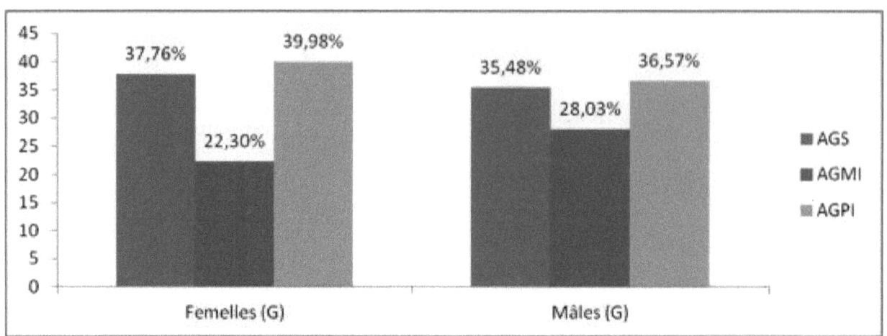

Figure 24: Variations in percentages of saturated (SFA), monounsaturated (MUFA) and polyunsaturated (PUFA) fatty acids in the gonads of the *Symphodus tinca* island population as a function of sex (month of March).

Figure 24 shows that the TFAs of female gonads are richer in SFAs and PUFAs than those of males, with approximate levels of 37.76% and 35.48% for SFAs, and 39.98% and 36.57% for PUFAs. Nevertheless, the MUFA content of male gonad lipids (28.03%) is higher than that of females (22.30%).

Table 18 shows variations in gonadal PUFA levels in the 3 populations studied.

Table 18: Comparison of polyunsaturated fatty acid (PUFA) percentages in the gonads of the three *Symphodus tinca* populations according to sex (month of March).

% TFA	Marine station	Lagoon plant	Island resort
% in PUFA from gonads females	58,25%	42,68%	39,98%
% in PUFA from male gonads	51,94%	59,30%	36,57%

The PUFA content of TFAs in female gonads is 39.88%, while in male gonads it is estimated at 36.57%. According to Table 18, these PUFA levels are the lowest in relation to other gonadal lipids in both marine and lagoon populations. Although, the lagoon males are characterized by: the richest gonads in PUFAs and the lowest in SFAs. This is shown by comparing their percentages with those of the other populations (Figures 18, 21 and 24).

A comparison of the (PUFA) percentages of the population's fillets and gonads is shown in Table 19. TFAs in the gonads of island individuals show PUFA levels of around 39.98% and 36.57% in male and female gonads respectively. These levels are higher than those found in fillets.

Table 19: Comparison of fillet and gonad (PUFA) percentages of the *Symphodus tinca* island population by sex (month of March)

% TFA	Females	Males
% of PUFAs in the gonads	39,98%	36,57%
% PUFAs in flesh	33,71%	30,33%

The results also showed that the lipids in the flesh of females are richer in PUFAs than those of males.

3-4-Comparison of major fatty acids of individuals (♂+♀) from the island population.

The three major GAs in the composition of TFAs in *Symphodus tinca* flesh are Populating the island coasts are C16:0, C18:1n9 and C20:4n6 (Table 20). For the entire population, palmitic acid is the most dominant, followed by ARA and, to a lesser extent, oleic acid.

Table 20: Variations in the percentages of major fatty acids in *Symphodus tinca* fillets from the island population (month of March).

% TFA	Females	Males
Major SFA (palmitic acid)	28,02%	24,76%
Major MUFA (oleic acid)	18,43%	21,12%
Major PUFAs (ARA)	11,67%	9,88%

The C16:0 and C20:4 n6 TFA content of female fillets is higher than that of male fillets. The levels are : 28.02% and 24.76% palmitic acid; 11.67% and 9.88% ARA, for female and male fillets respectively.

The oleic acid content of these fillets was 21.12%, higher than that of the females (18.43%).

According to Table 21, the SFAs and MUFAs in livers have the same MFAs as in flesh, but the major PUFA in livers is EPA (instead of ARA).

Palmitic acid has the highest level, oleic acid the highest, and EPA the lowest.

Table 21: Variations in the percentages of major fatty acids in the livers of the *Symphodus tinca* island population (month of March).

% TFA	Females	Males
Major SFA (palmitic acid)	26,55%	25,70%
Major MUFA (oleic acid)	18,27%	20,18%
Major PUFA (EPA)	11,30%	10,22 %

For the majority fatty acids making up liver lipids, they are the same for the SFAs and MUFAs contained in the flesh, while C20:4 n-6 will be substituted by C20:5 n-3, (table 20 and 21).

According to Table 22, the gonadal MGAs of the island population are: C16:0, C18:9, and C22:6, n-3. C16:0 always has higher levels than C18:n9, while C22:6 n-3 has the lowest.

Table 22: Variations in major fatty acid percentages in the gonads of the island population as a function of sex (month of March).

% TFA	Females	Males
Major SFA (palmitic acid)	26,95%	26,48%
Major MUFA (oleic acid)	15,77%	20,96%
Major PUFA (DHA)	13,12%	9,15%

In terms of lipid profiles for male and female gonads, C16:0 and C22:6n-3 levels are similar, but slightly higher in female gonads. These levels are : 26.55% and 25.70% for C16:0; 13.12% and 9.15% for C22:6n-3.

On the contrary, the C18:n9 content of male gonads is higher than that of ovaries, averaging 20.18% for male gonads and 15.77% for female gonads.

Tables 20, 21 and 22 show that MFAs of SFAs and PUFAs of the 3 organs are higher in females than in males. Males, on the other hand, have the highest levels of (C18:1 n9) in all three organs.

III-4-Study of variations in omega-3 and omega-6 content (month of March)

The results in Tables 23,24...28 show omega-3 and omega-6 levels for each organ and for the 3 populations. It can be seen that omega-3s have higher levels than omega-6s.

4-1-Study of the omega-3 and omega-6 composition of the *Symphodus tinca* marine population

Table 23 shows the variation in omega-3 levels in each organ of the marine individuals. It can be seen that the fatty acid composition of the three female organs contains more omega-3 than that of the males, with a maximum recorded for the flesh of female specimens estimated at 37.47% and a minimum of around 22.04% for the TFAs of male livers.

Table 23: Variations in omega-3 percentages in the marine population of *Symphodus tinca* (month of March).

% TFA	Chair	Livers	Gonads
Females	37,47%	28,92%	36,21%
Males	32,52%	22,04%	35,36%

For male marine specimens, the organ richest in omega 3 is the gonad, while for females, it's the flesh.

The lowest omega-3 content was recorded in livers, with 28.92% for female livers, compared with 22.04% for male liver lipids.

The arrangement of omega-6 levels is similar to that of omega-3s, i.e. at the level of the three organs: females have more omega-6s than males (Table 24), with an estimated maximum of 18.11% for female fillets and a minimum for male liver TFAs, whose content is around 13.11%.

Table 24: Variations in omega-6 percentages in the marine population of *Symphodus tinca* (month of March).

% TFA	Chair	Livers	Gonads
Females	18,11%	13,36%	18,04%

| Males | 15,33% | 13,11% | 16,57% |

4-2-Comparison of the omega-3 and omega-6 composition of the *Symphodus tinca* population occupying lagoon waters

A comparison of omega-3 levels in each organ of lagoon-dwelling individuals shows that these levels vary according to sex. TFA levels in the flesh and gonads of male specimens are higher than those of female specimens. These levels are : 37.65% and 32.75% in the flesh; 41.01% and 28.53% in the gonads of males and females respectively (Table 25).

Female livers, on the other hand, have a higher omega-3 content, around 25.40%, than male livers, estimated at 19.38%. Based on these levels, it can be concluded that the liver always has the lowest levels of omega-3 compared to the flesh and gonads.

Table 25: Variations in omega-3 percentages in the lagoon population of *Symphodus tinca* (month of March).

% TFA	Chair	Livers	Gonads
Females	32,75%	25,40%	28,53%
Males	37,65%	19,38%	42,01%

The arrangement of omega-6 TFAs in the three organs is similar to that of omega-3s, with male flesh and gonad TFAs marking higher levels than those of coexisting females (Table 26). These levels are 37.65% and 32.75% for flesh TFAs, and 41.01% and 28.53% for gonad TFAs in males and females respectively.

Table 26: Variations in omega-6 percentages in the lagoon population of *Symphodus tinca* (month of March).

% TFA	Chair	Livers	Gonads
Females	14,24%	13,20%	14,14%
Males	14,43%	11,00%	15,74%

Similarly, female livers have higher levels of omega 6 (25.40%) than male livers (19.38%). Thus, liver TFAs always have the lowest levels of omega 3 compared to flesh and gonads.

4-3-Comparison of omega-3 and omega-6 composition between males and females of *Symphodus tinca* on island coasts (March)

A comparative study of omega-3 levels in each organ of marine individuals shows variations according to sex (Table 27). In all three organs, females have higher levels of omega-3 than males. The maximum percentage recorded for female gonads, estimated at 27.79%, and a minimum of around 16.49% for TFAs in male flesh.

Table 27: Variations in omega-3 percentages in the island population of *Symphodus tinca* by sex (month of March).

% TFA	Chair	Livers	Gonads
Females	17,85%	21,35%	27,79%
Males	16,49%	19,28%	20,29%

Table 27 shows that for all male and female specimens, the organs with the highest omega-3 content are the gonads, with levels of 20.29% and 27.79% for male and female gonads respectively. The lowest omega-3 content was recorded in the flesh, with proportions of 17.85% and 16.49% for females and males respectively. This result differs from that of the other two populations, whose minimum omega-3 levels are recorded in the livers.

The arrangement of omega-6 levels differs from that of omega-3s, with male liver and gonad TFAs accumulating more omega-6s than those of females (Table 28). Rates are 12.98% and 9.31% for livers, and 14.18% and 12.21% for gonads, for males and females respectively.

Table 28: Variations in omega-6 percentages in the island population of *Symphodus tinca* (month of March)

% TFA	Chair	Livers	Gonads
Females	16,51%	9,31%	12,21%
Males	13,83%	12,98%	14,18%

Males, on the other hand, are characterized by the richest omega-6 fillets, with levels of around 16.51% for male livers and 13.83% for female livers.

III-5-Study of variations in the ω6 ω3 ratio in the three organs of the three populations

The nine Lnc PUFAs identified are grouped together as omega-6 and omega-3, and a ratio is used to compare their levels and determine which is the largest and smallest family of PUFAs:

$$R = \omega 6 / \omega 3$$

This ratio is determined for each fatty acid extract of the 3 organs.

If: R < 1, APGI are essentially made up of ω6.

1 < R, APGI consists mainly of ω3.

By establishing these ratios for all fatty acid extracts, we note that they are always less than 1 (tables 29, 30 and 31), so APGI is made up mainly of ω3 and to a lesser extent of ω6. This quotient oscillates between a maximum recorded for the flesh of island females (0.9604) and a minimum of 0.3745 for the flesh of lagoon males.

Table 29 shows variations in the omega-6/omega-3 ratio for the three *Symphodus tina* organs at the marine station.

Table 29: Variation in the omega-6 to omega-3 ratio in three *Symphodus tina* organs at the marine station (month of March).

ω6/ω3	Chair	Livers	Gonads
Females	0,5070	0,4713	0,4035
Males	0,4805	0,6019	0,4697

Based on these results, comparing the (R)s of livers, fillets and gonads, we note that the richest PUFAs in ω 3 are those of the gonads (the highest ratios), followed by muscles and, to a lesser extent, livers.

Table 30 shows variations in the ω6 ω3 ratio for the three *Symphodus tina* organs at the lagoon station.

Table 30: Variation in the omega-6 omega-3 ratio in three *symphodus tina* organs at the lagoon station (month of March).

ω6/ ω3	Chair	Livers	Gonads
Females	0,4368	0,5187	0,5165

	0,3745	0,5699	0,3784
Males			

Based on these results, we deduce that the flesh of individuals from the lagoon environment is the richest in ω3, with the lowest ratios of 0.4368 (for female fillets) and 0.3745 (for male fillets). Gonads and liver are less rich in ω3, with higher ratios (R) of 0.5187 and 0.5699 for female and male livers respectively.

Table 31 shows the variations in the ω6 ω3 ratio in the three organs of the island population.

Table 31: Variation in the omega-6 to omega-3 ratio in three *symphodus tina* organs at the island station (month of March).

ω6/ ω3	Chair	Livers	Gonads
Females	0,9604	0,6971	0,4540
Males	0,8554	0,7133	0 ,7003

For the island population, gonadal lipids are the richest in ω 3, (the lowest ratio), with liver lipids in lower proportions and flesh lipids at a lower level (Table 31). This can be explained by the mobilization of energy for the gonads, which are the most mature of all the stations. This transfer of lipids takes place from the flesh and liver to the gonads.

Tables 29 and 31 show that in the 2 environments (marine and island), the (R)s of the male gonads are always higher than those of the ovaries. Conversely, for lagoon specimens, the (R)s of ovaries and testes show that the latter are the richest in ω 3.

Based on these tables, it can be seen that the (R) ratios for hepatic lipids in males are higher than those in females for each of the 3 populations. These ratios are: (0.4713 and 0.6019); (0.5187 and 0.5699); (0.6971 and 0.7133), successively for females and males; marine, lagoon and island. Thus, the TFAs of female livers are richer in Ln3 PUFAs than those of males.

Within each population studied, the (R)s of male fillets are always lower than those of females: (0.5070 and 0.48050), (0.4368 and 0.3745) and (0.9604 and 0.8554); respectively for females and males: marine, lagoon and island. Thus, male flesh tends to accumulate ω3 more than ω6.

Finally, by comparing these ratios for all the populations studied, we note that the TFAs in the flesh of lagoon individuals are the richest in ω 3, followed by those of marine individuals, and lower for fillets from island individuals.

Discussion

S. tinca plays an important ecological role due to its ability to withstand wide variations in environmental physicochemical parameters, which explains its presence in various aquatic environments.

The first part of this study aimed to determine the morphometry of S. *tinca such as* gonado-somatic and hepato-somatic factor. The second part was designed to determine the total lipid content and lipid composition of the flesh, livers and gonads of three populations of *S. tinca* occupying different biotopes.

During the sampling period (March), morphometric studies of three *S. tinca* populations showed that RGS were always above 1%. This is in line with the work of (Ouannes-Ghorbel et al., 2002), which shows that *Symphodus tinca* is maturing.

The results also showed that the gonado-somatic ratio evolves inversely in comparison with the hepato-somatic ratio.

Thus, (Ouannes-Ghorbel et al., 2002), who show that RGS increases at the expense of RHS, the liver mass of males and females decreases while their gonads grow, due to the fact that the maturation of *S. tinca* gonads stores the reserves necessary for reproduction in the liver.

This study shows that lipid levels vary according to organ (flesh, liver and gonads), sex and environment (marine, island and lagoon).

The present study shows that fish in the 3 environments studied are characterized by an average fat content in the flesh ranging from a minimum of 2.26 g/100 g MF for male lagoon specimens to a maximum of 3.19 g/100 g MF for male island specimens. These results indicate that the species is classified as semi-fat to lean, with fillet lipids ranging from 1 to 10% (Ackman and Sipos, 1964).

 On the other hand, for all the populations studied, the lipid contents determined in muscle (M), liver (F) and gonads (G) allow us to conclude that: lipid accumulation occurs primarily in the liver, secondarily in the gonads and finally in the muscle, which is typical for lean fish.

Consequently, almost all lipids are stored in the form of triglycerides in the hepatocytes to provide a source of energy that can be used when the food is not sufficiently supplied. Energy requirements are very high, especially during the breeding (gamete and egg production) and migration periods.

Ovogenesis, like spermatogenesis, in fish requires significant energy input, which so-called "lean" species store in the liver, mainly in the form of lipids (Hoar, 1957; Bertin, 1958). This mobilization is short-term from the storage site, but long-term for lipids in other tissues (fillets).

Generally speaking, fish have a seasonal reproductive cycle involving maturation and development phases. During this cycle, and more specifically at the gonadal maturation stage, which is preceded by a period of energy storage in the form of mobilization lipids, energy is transferred from the deposit sites to the liver, where it is mobilized and then transported to the gonads. Transfer takes place from the deposit sites to the liver, where they are mobilized and transported to the gonads (Craig et al., 2000), which is why gonads have higher lipid contents than flesh.

GPC analysis of fatty acid methyl esters from different organs identified 15 fatty acids, grouped into saturated (SFA), monounsaturated (MUFA) and polyunsaturated (PUFA) fatty acids. As a result, the fatty acid profile of different organs varies according to their biotope.

Crenilabra from the Ghar Elmelh lagoon have the highest PUFA content compared to other populations, which may be explained by differences in salinity, diet, water temperature or location (Guler et al., 2007). An increase in PUFA content and a decrease in saturated PUFAs in membrane phospholipids are observed when fish move to the saltier environment (Greene and Selivonchick, 1987).

Screening of the lipid profile of the flesh from different populations shows that island individuals contain the highest levels of SFA. These high levels of PUFAs are essentially explained by the degree of maturity, which is more advanced than in the other 2 batches derived from the northern coasts of Tunisia. Thus, PUFAs are mainly stored during the sexual resting period, when food is abundant, and then mobilized during maturation. However, during the reproductive phase, PUFAs are mobilized and depleted in order to complete gonad maturation (Guler et al., 2007).

For all three populations, quantification of SFAs revealed that the liver contained the highest percentages and levels. These high SFA levels can be explained by the fact that this organ is the site of numerous metabolic reactions affecting fatty acids. In fact, the main site of new FA synthesis in teleosts is the liver (Hendeson et al., 1982; Sargent et al., 1989).

More importantly, the lipid profiles of the flesh from the 3 populations show that palmitic acid is the most important fatty acid in this fraction, followed by C18:0 and

C14:0. The dominance of palmitic acid in fish lipids has been reported by other authors (Bandarra et al, 2001; Osman et al, 2001; Aidos et al, 2002; Passi et al, 2002). This percentage of fatty acids seems to be fairly constant throughout the year. Ackman and Eaton (1966) reported similar data in a study with herring and concluded that the content of this compound does not appear to be influenced by diet.

 This study reveals that the edible tissues of peacock wrasse are very rich in omega-3 fatty acids, but contain low levels of omega-6 fatty acids. These results are similar to those of Oksuz et al (2011). Thus, it is important to consume more fish rich in ω3 fatty acids and low in ω6 fatty acids (Sargent et al., 1997). In fact, ω3 fatty acids are nutrients sought after and recommended by nutritionists because of their notable benefits, especially through their preventive action in the development of several diseases.

The results in Tables 27,28...32 show the levels of omega-3 and omega-6, for each organ and for the 3 populations. It can be seen that: omega-3s have higher levels than omega-6s. These results are in line with the work of Kaya et al.(2017).

Anisi, the richness in ω3 fatty acids then became a selling point for these types of products. In this context, an approach covering the quality and value-added of fish with a low market value had to be developed.

Conclusion

In this study, we will examine the morphometry, total lipid content and fatty acid composition of the flesh, livers and gonads of 3 populations of *Symphodus tinca* from 3 different biotopes. Our study also focuses on the valorization of Symphodus tinca based on the identification and quantification of fatty acids such as PUFAs (ω3 and ω6), known for their roles in various physiological functions and the prevention of various pathologies.

The morphometric study revealed that the hepato-somatic ratio evolves inversely to the gonado-somatic ratio, with the latter always showing a value greater than 1, demonstrating that the species is in the reproductive period.

Determination of total lipid content revealed that the liver represented the highest levels, followed by the gonads and, to a lesser extent, the flesh.

Gas chromatography analysis of organ lipid extracts identified 19 fatty acids, with significant variations in the percentage of saturated, monounsaturated and polyunsaturated fatty acids at organ level and according to sampling station.

Quantification of polyunsaturated fatty acids reveals that omega-3 levels are higher than omega-6 levels in various organs.

Thus, in terms of flesh, the lagoon population is the richest in ω 3, followed by marine fish and, to a lesser extent, island individuals.

References

Abdul Malak, D., Livingstone - Suzanne, R., Pollard, D., Polidoro, B., Cuttelod, A., Bariche, M., Bilecenoglo, M., Carppenter, K.E., Colette, B., Francour, P., Goren, M., Kara, M., Massuti, E., Papaconstantino, C. and Tunesi, L. 2011. Overview of the conservation status of marine fishes occurring in the Mediterranean Sea. *Gland, Switzerland and Malaga, Spain: IUCN. Vii,* 61 pp.

Ackman, R.G. and Sipos, J.C. 1964. Application of specific response factors in the gas chromatographic analysis of methyl esters of fatty acids with flame ionization detectors. *Journal of the American Oil Chemists Society, 41*(5), pp.377-378.

Ackman, R.G. and Eaton, C.A. 1966. Some commercial Atlantic herring oils; fatty acid composition. *Journal of the Fisheries Board of Canada, 23*(7), pp.991-1006.

Ackman, R. 1980. Fish lipids. In Advances in Fish science and Technology (J. J. Connell ed.)Fishing News Books Ltd, Surrey (England).

Ackman, R. 1995. Composition and nutritional value of fish and shellfish lipids. In: Fish and Fisheries Products (Ed: Ruiter, A*.) Cab International, Wallingford, Oxon,* UK.p.117-156.

Adams, P., Lawson, S., Sanigorski, A and Sinclair, A. 1996. Arachidonic acid to eicosapentaenoic acid ratio in blood correlates positively with clinical symptoms of depression. *Lipids.*31:157-161.

Aksnes, A.T., Hjertnes and Opstvedt. J. 1996. Effects of dietary protein level on growth and carcass composition in Atlantic halibut (*Hippoglossus hippoglossus L*). *Aquaculture,* 145:225-233.

Aidos, I., Van Der Padt, A., Luten, J.B. and Boom, R.M. 2002 . Seasonal Changes in Crude and Lipid Composition of Herring Fillets, By-products, and Respective Produced Oils. *J. Agr. Food Chem.* 16: 4589-4599.

Andrews, J., Murray, W., and Davis, M. 1978. The influence of dietary fat levels and environmental temperature on the digestible energy and absorbability of animal fat in catfish diets. *J. Nutr.* 108: 749-752.

Ando, S., Mori, Y., Nakamura, K. and Sugawara, A. 1993. Characteristics of lipid accumulation types in five species of fish. *Nip. Suis. Gak.* 59: 1559-1564.

Arts, M. and Kohler, C. 2009. Health and condition in fish: the influence of lipids on membrane competence and immune response. In: Arts M, Brett M, Kainz M (eds) *Lipids in Aquatic Ecosystems. Springer New York*, p.237-255.

Augier, H. and Boudouresque, F. 1979. Première observation sur l'herbier de posidonie et le détritique côtier de l'ile du levant (Méditerrané, France) *Trav. Sci. Parc nation port- crocs*, 5 :141-153.

Azouz, A. 1973. Trawlable bottoms in the northern region of Tunisia. 1: Physical framework of the northern coasts of Tunisia. *Bull. Inst. Océanogr. Pêche. Salammbô.* 2 (4): 473-564.

Balk, E., Lichtenstein, A., Chung, M., Kupelnick, B., Chew,P. and Lau, J. 2006. Effects of omega-3 fatty acids on coronary restenosis, intima-media thickness, and exercise tolerance: a systematic review. *Atherosclerosis,* 184: 237-246.

Bandarra, N.M., Batista, I., Nunes, M.L. and Empis, J.M. 2001. Seasonal Variation in the Chemical composition of Horse Mackerel (*Trachurus trachurus*).*Eur. Food Res. Technol.* 212: 535-539.

Bauchot, O.M.l. and Pras, A. 1980. Guide de poissons marins de l'Europe, les guides de naturaliste Edit., *De la chaux et Nestlé, Lausanne,* 427p.

Bell, J.D. and Harmelin-Vivien, M.L. 1982. Fish fauna of French Mediterranean Posidonia oceanica seagrass meadows. I. Community structure. Tethys, 10: 337-347.

Bell, M., Henderson, R. and Sargent, J. 1986. The role of polyunsaturated fatty acids in fish.*Comp. Biochem. Physiol.* 83(B): 711-719.

Ben Slama, S., Menif, D. and Ben Hassine, O.K. 2006. Diet of Labrus merula (Labridae) from the northern coasts of Tunisia. *Rencontres de l'Ichtyologie en France No3, Paris, France* 6 :175-180.

Ben Othman, S. 1973. Le sud Tunisien (golfe de gabes: hydrologie et sédimentologie, flore et faune. *Thèse 3ème Cycle, Univ. Tunis,* 166 p.

Berenson, G.S., Srinivasan, S.R., Bao, W., Newman, W.P., Tracy, R.E. and Wattigney, W.A. 1998. Association between multiple cardiovascular risk factors and atherosclerosis in children and young adults. The Bogalusa Heart Study. *New England Journal of Medicine.* 338:1650-1656.

Berg, L. 1985. The present and fossil fish-like system and fish. Berlin: *VEB German Academic Publishers.*

Bertin, L. 1958. Organs of aquatic respiration. In: GRASSÉ PP (Ed), Traité de Zoologie, *Tome XIII (II), Masson& Cie, Paris:* 1303-1341

Bligh, E. and Dyer, W. 1959. A rapid method of total lipid extraction and purification. *Can. J.Biochem. Physiol,* 37: 911-917.

Bourquard C., 1985. Structure and mechanisms of establishment, maintenance and evolution of ichthyic populations in the Gulf of Lion. Doctoral thesis, 312 p. Univ. Montpellier 2.

Bradai, M.N. 2000. Diversité du peuplement ichtyque et contribution à la connaissance des sparidés du golfe de Gabès. *Thèse Doct d'État. Sciences Nat, Fac. SCI. Sfax,* 600 p.

Brauge, C., Corraze, G., and Médale, F. 1995. Effects of dietary levels of carbohydrate and lipid on glucose oxidation and lip genesis from glucose in rainbow trout, *Oncorhynchus mykiss,* reared in freshwater and seawater. *Comp. Biochem. Physiol, INRA,* P. 117-124.

Brett, M., Muller-Navara, D. and Perrson, J. 2009. Crustacean zooplankton fatty acid composition. In: Arts M, Brett M, Kainz M (eds). *Lipids in aquatic ecosystems. Springer, New York,* p.115-146.

Brett, J.R., 1979. Environmental factors and growth. In *Fish physiology* (Vol. 8, pp. 599-675). Academic press.

Brockerhoff, H., Ackman, R., and Hoyle, R. 1963. Specific distribution of fatty acids in marine lipids. *Arch Biochem Biophys,* 100: 9-12.

Budge, S.M., Iverson, S.J. and Koopman, H.N. 2006. Studying trophic ecology in marine ecosystems using fatty acids: a primer on analysis and interpretation. *Mar. Mamm. Sci.* 22, 759-801.

Calder, P. 2009. Polyunsaturated fatty acids and inflammatory processes: New twists in an old tale. *Biochemistry,* 91: 791-795.

Camchi, M., Fowler, S.L., Musick, J.A.,Brautigam, A. and Fordham, S.V. 1998. Sharks and their relatives-Ecology and conservation. IUCN/SSC *Shark Specialist Group. UICN, Cland, Switzerland and Cambridge, UK,* 39 pp.

Cailliet, G.M., Musick, J.A., Simpfendorfer, C.A. and Steven, J.D. 2005. Ecology and Life History Carachteristics of Chondrichthyan Fish. *In*: Fowler, S.L. ; Cavanagh, R.D. ; Camchi, M. ; Burgess, G.H. ; Cailliet, G.M. ; Forhham, S.V. ; Simpfendorfer, C.A and Musick, J.A. 2005. *Sharks, rays and Chimaeras: The statue of the condrichthyan Fishes. IUCN/SSC Shark Specialist Group.UICN, Gland, Switzerland and Cambridge, UK,* 12-18 pp.

Canler, J. 2001. Performances des systèmes de traitement biologique aérobie des greisses - Graisses issus des dégraisseurs de stations épuration traitant des effluents à dominante domestique. *Document technique FNDAE* n°24, 62 p.

Capanni, M., Calella, F., Biagini, M.R., Genise, S., Raimondi, L. and Bedogni, G. 2006. Prolonged n-3 polyunsaturated fatty acid supplementation ameliorates hepatic steatosis in patients with non-alcoholic fatty liver disease: a pilot study. *Aliment Pharmacol Ther.* 23:1143-1151.

Carlson, S.E., Cooke, R.J., Rhodes, P.G., Peeples, J.M. and Werkman, S.H. 1992. Effect of vegetable and marine oils in preterm infant formulas on blood arachidonic and docosahexaenoic acids. *The Journal of pediatrics, 120*(4), pp.S159-S167.

Casabinanca, M., De Kiener, A. and Huve, H. 1972. Biotopes and biocenosis of Corsican brackish ponds: *Biguglia, Diana, Urbino, Palo*. Milieu vol XXIII, fasc. ser. p. 187-227.

Cecchi, G., Biasini, S. and Castano, J. 1985. Rapid methanolysis of oils in medium solvent. *Revue Francaise des Corps Gras*, 32: 163-164.

Chavarro, J., Stampfer, M.L.I.H., Campos, H., Kurth, T. and Ma, J. 2007. A Prospective study of polyunsaturated fatty acid levels in blood and prostate cancer risk. *Cancer Epidemio Biomarkers Prev,* 16:1364-1370.

Cho, C. and Bureau, D. 2001. A review of diet formulation strategies and feeding systems to reduce excretory and feed wastes in aquaculture. *Aquacult. Res.* 32: 349-360.

Chong, E.W. and Kreis, A.J. 2008. Dietary omega-3 fatty acid and fish intake in the primary prevention of age-related macular degeneration: a systematic review and met analysis. *Archives of Ophthalmology,* 126(6): 826-833.

Corraze, G. and Kaushik, S.J. 1999. Lipids in marine and freshwater fish. *OCL,* 6: 111-115.

Cowey, C. 1993. Some effects of nutrition on flesh quality of cultured fish. In: Fish Nutrition in Practice (Eds: Kaushik, S.J., Luquet, P.) Proc. IV Int. Symp. Fish Nutrition and feeding, Les colloques INRA, n°61, Editions INRA, Paris, France, .227-236.

Craig, S.R., Mackenzie, D.S., Jones, G. and Catlin, D.M. III. 2000. Seasonal changes in the reproductive conditions on body composition of three rearing red drum, *"Sciaenops ocellatus". Aquaculture:* 190(1-2)89-102.

Crockett, E. and Sidell, D. 1993. Peroxisomal β-Oxidation Is a Significant Pathway for Catabolism of Fatty Acids in a Marine Teleost. *Am J Physiol.* 264: 1004-1009.

Cuijpers, P. and Smith, F. 2002. Excess mortality in depression: a meta-analysis of community studies. *J Affect Discord*; 72:227-236.

Dabrowski, K. and Guderley, H. 2002. Intermediary metabolism In: Fish nutrition (Eds: Halver, J.E., and Hardy, R.W.) *.Academic press,* pp. 309-365.

Denton, J. and Yousef, M. 1976. Body composition and organ weights of rainbow trout, *Salmo gairdneri. J Fish Biol,* 8:489-499

Dickinson, H.O., Mason, J.M., Nicolson, D.J., Campbell, F., Beyer, F.R., Cook, J.V., Williams, B. and Ford, G.A. 2006. Lifestyle interventions to reduce raised blood pressure: a systematic review of randomized controlled trials. *Journal of hypertension*, *24*(2), pp.215-233.

Dieuzeide, R. 1955. Catalog des poissons des cotes algériennes (III. Osteopterygii). *Bull. Stat. Aqui. Cast. 6*, pp.1-384.

Eldho, N., Feller, S., Tristram-Nagle, S., Polozov, IV. and Gawrisch, K. 2003. Polyunsaturated docosahexaenoic vs docosapentaenoic acid - Differences in lipid matrix properties from the loss of one double bond. *Journal of the American Chemical Society* 125: 6409-6421.

Entressangles, B., Pasero, L., Savary, P., Sarda, L. and Desnuelle, P. 1961. Influence of the nature of chains on the rate of their hydrolysis by pancreatic lipase. *Bull Soc Chim Biol*, 43:581-91.

FAO.2006. The State of World Fisheries and Aquaculture 2006. 180 pp.

Fauconneau, B., Andre, S., Chmaitilly, H., Le Bail, P., Krieg, F. and Kaushik, S. 1997. Control of skeletal muscle fibers and adipose cells size in the flesh of rainbow trout. *J Fish Biol,* 50: 296-314.

Faust, I., Johnson, P., Stern, J. and Hirsch, J. 1978. Diet-induced adipocyte number increase in adult rats: a new model of adiposity. *Am J Physiol,* 235: 275- 286.

Faust, J.M. and Miller, W.H. 1981. Hyperplasic growth of adipose tissue in obesity. In: The Adipocyte and obesity: Cellular and Molecular Mechanism (Eds: Angel, A., Wollenberg, C.H., and Roncari D.A.K.) *Raven Press, New York, USA*, pp.41-51.

Fischer, W., Bauchot, M. and Schneider, M. 1987. FAO Species Identification Sheets for Mediterranean and Black Sea Fisheries "Revision". Fishing Area 37. *Vertebrates. Rome FAO*, 2: 761-1530.

Fluckiger, F. 1981. Fish cleaning in the Mediterranean by *Crenilabrus melano-cercus*. Rapp. Comm. Int. Mer Medit, 27: 5.

Folch, J., Lees, M. and Sloan Stanley, G. 1957. A simple method for the isolation and purification of total lipids from animal tissues. *J Biol Chem*, 226 :497-509.

Fonteneau, 1995. Mediterranean deep-sea pelagics: fishing, research and resource management - Current situation and outlook in: *La pêche maritime (May-June 1995)*

Frasure-Smith, N., Lespérance, F. and Julien, P. 2004. Major depression is associated with lower omega-3 fatty acid levels in patients with recent acute coronary syndromes. *Biological psychiatry, 55*(9), pp.891-896.

Gélineau, A., Corraze, G., Boujard, T., Larroquet, L. and Kaushik, S. 2001. Relation between dietary lipid level and voluntary feed intake, growth, nutrient gain, lipid deposition and hepatic lipogenesis in rainbow trout. *Reproduction Nutrition Development, 41*(6), pp.487-503.

Geleijnse, J., Giltay, E., Grobbee, D., Donders, A. and Kok, F. 2002. Blood pressure response to fish oil supplementation: Metaregression analysis of randomized trials. *J Hypertens,* 20: 1493-1499.

Greer-Walker, M. 1970. Growth and development of the skeletal muscle fibers of the cod (*Gadus morhua* L.). *J Cons Perm Int Pou Mer.* 33: 228-244.

Greene, D. and Selivonchick, D. 1987. Lipid metabolism in fish. *Prog. Lip. Res.* 26: 53-85.

Guler, G.O., Aktumsek, A., Citil, O.B., Arslan, A. and Torlak, E. 2007. Seasonal variations on total fatty acid composition of fillets of zander (Sander lucioperca) in Beysehir Lake (Turkey). *Food Chemistry, 103*(4), pp.1241-1246.

Hamazaki, T., Sawazaki, S., Nagao,Y., Kuwamori, T., Yazawa, K., Mizushima, Y. and Kobayashi, M. 1998. Docosahexaenoic acid does not affect aggression of normal volunteers under nonstressful conditions. A randomized, placebo-controlled, double-blind study. *Lipids* 33: 663-667.

Hamazaki, K., Itomura, M., Huan, M., Nishizawa, H., Sawazaki, S., Tanouchi, M., Watanabe, S., Hamazaki, T., Terasawa, K. and Yazawa, K. 2005. Effect of omega-3 fatty acid-containing phospholipids on blood catecholamine concentrations in healthy volunteers: a randomized, placebo controlled, double-blind trial. *Nutrition,* 21: 705-710.

Harmelin-viven, M. 1982. Ichtyofaune des herbiers de posidonies de parc national de port Crocs: I Composition et variation spatio-temporelles, *Trav. Sci. Parc nation. Port-Cross*, 8: 69-92.

Hattori, T., Adachi, K. and Shizuri, Y. 1998. New ceramide from marine sponge Halicona koremella and related compounds as antifouling substances against macroalgae. *J Nat Prod*, 61: 823-826.

Hedelin, M., Chang, E., Wiklund, F., Bellocco, R., Klint, A., Adolfsson, J., Shahedi, K.Xu.J., Adami, H., Gronberg, H. and Balter, K. 2007. Association of frequent consumption of fattyfish with prostate cancer risk is modified by COX-2 polymorphism. *Int J Cancer,* 120: 398-405.

Hellio, C., Bremer, G., Pons, A.M., Le Gal, Y. and Bourgougnon, N. 2000. Inhibition of the development of microorganisms (bacteria and fungi) by extracts of marine algae from Brittany, France. *Appl Microbiol Biotechnol*, 54: 543-549.

Hemre, G. and Sandnes, K. 1999. Effect of dietary lipid level on muscle composition in Atlantic salmon *Salmo salar. Aquacult.Nut*. 5: 9-16.

Hemre, G., Mommsen, T. and Krogdahl, A. 2002. Carbohydrates in fish nutrition: effects on growth, glucose metabolism and hepatic enzymes, *Aquaculture. Nutr.* 8, 175-194.

Henderson, R.J., Sargent, J.R. and Pirie, B.J.S. 1982. Peroxisomal oxidation of fatty acids in livers of rainbow trout *(Salmo gairdneri)* fed diets of marine zooplankton. Comparative Biochemistry and Physiology, 73B, 565-57.

Henderson, L. and Tocher, D. 1987. The lipid composition and biochemistry of freshwater fishes, *Prog. Lipid Res.* 26: 281-347.

Henderson, R., Millar, R. and Sargent, J. 1995. Effect of growth temperature on the positional distribution of eicosapentaenoic acid and transhexadecenoic acid in the phospholipids of a *Vibrio* species of bacterium. *Lipids* 30: 181-185.

Hennen, G. 1995. Biochimie 1er cycle. *Dunod:* Paris, 436 p.

Hibbeln, J. 1998. Fish composition and major depression. *Lancet* 351:1213.

Hillestad, M. and Johnsen, F. 1994. High energy/ low protein diets for Atlantic salmon: effects on growth, nutrient retention and slaughter quality. *Aquaculture,* 124: 109-112.

Hirose, K., Takezaki, T., Hamajima, N., Miura, S. and Tajima, K. 2003. Dietary factors protective against breast cancer in Japanese premenopausal and postmenopausal women. *Int J Cancer,* 107: 276-282

Hoar, W.S. 1957. The gonads and reproduction - in the physiology of fishes L Metabolism. - Brown M.E Ed. Ac. Press 287-321.

Hoffman, S.G., Schildhauer, M.P. and Warner, R.R. 1985. The cost of changing sex and the ontogeny of males under contest competition for mates. *Evolution,* 39: 915-922.

Holmström, C. and Kjelleberg, S. 1999. Marine Pseudoalteromonas species are associated with higher organisms and produce biologically active extracellular agents. *FEMS. Microbiol. Ecol.* 30: 285-293.

Hughes, T. A., Heimberg, M., Wang, X., Wilcox, H., Hughes, S. M., Tolley, E. A., Desiderio, D. M., and Dalton, J. T. 1996. Comparative lipoprotein metabolism of myristate, palmitate, and stearate in normolipidemic men. *Metabolism,* 45, 1108-18.

International union of pure and applied chemistry and international union of biochemistry commission on biochemical nomenclature,1978. The nomenclature of lipids. J Lipid Res, 19: 114-129.

Jobling, M., Koskela, J. and Savolainen, R. 1998. Influence of dietary fat level and increased adiposity on growth and fat deposition in rainbow trout, *Oncorhynchusmykiss* (Walbaum). *Aquacult. Res.* 29: 601-607.

Johansson, L., Kiessling, A., Kiessling, K. and Berglund, L. 2000. Effects of altered ration levels on sensory characteristics, lipid content and fatty acid composition of rainbow trout (*Oncorhynchus mykiss*), *Food Qual Prefer.*11: 247-254.

Kapoor, B., Smith, H. and Verighina, I. 1975. The alimentary canaland digestion in teleosts. *Adv. Mar. Biol.* 13:109-239.

Kaushik, S. and Oliva-Teles, A. 1985. Effect of digestible energy on nitrogen and energy balance in rainbow trout. *Aquaculture,* 50: 89-101.

Kaushik, S. 1986. Environmental effects on feed utilization. *Fish Physiol Biochem*, 2: 131-140.

Kaushik, S. 1997. Nutrition-feeding and body composition in fish. *Cah .Nutr.Diet.*32: 100-106.

Kaya, G. and Türkoğlu, S. 2017. Analysis of certain fatty acids and toxic metal bioaccumulation in various tissues of three fish species that are consumed by Turkish people. *Environmental Science and Pollution Research, 24*, pp.9495-9505.

Koven, W., Van Anholt, R., Lutzky, S., Atia, I.B., Nixon, O., Ron, B. and Tandler, A. 2003. The effect of dietary arachidonic acid on growth, survival, and cortisol levels in different-age gilthead seabream larvae (Sparus auratus) exposed to handling or daily salinity change. *Aquaculture, 228*(1-4), pp.307-320.

Kramer, J., Parodi, P., Jensen, R., Mossobam, M., Yurawecz, M. and Adolf, R. 1998. Rumenic acid: a proposed name for the major conjugated linoleic acid isomer found in natural products. *Lipids*, 33: 835

Le Bris, S., Pean, M. and Guichard, B. in: DORIS, 28/04/2013: *Symphodus Tinca* (Linnaeus, 1758), http://doris.ffessm.fr/fiche2.asp?fiche_numero=604.

Le Grand, P. and Rioux, V. 2010. The complex and important cellular and metabolic functions of saturated fatty acids. *Lipids, 45*, 941-6.

Lee, T.H., Hoover, R.L., Williams, J.D., Sperling R.I., Ravalese J., 3rd, Spur, B.W., Robinson, D.R., Corey, E.J., Lewis, R.A. and Austen, K.F. 1985. Effect of dietary enrichment with eicosapentaenoic and docosahexaenoic acids on in vitro neutrophil and monocyte leukotriene generation and neutrophil function. *N Engl J Med*; 312(19):1217-1224.

LeTourneur Y., 1991. Changes in the fish population of the Saint-Pierre reef flat (Reunion Island, Indian Ocean) following the passage of cyclone Firinga. *Cybium*, 15: 159-170.

Lie, Ø., Waagbø, R. and Sandnes, K. 1988. Growth and chemical composition of adult Atlantic salmon (*Salmo salar*) fed dry and silage-based diet. *Aquaculture,* 69: 343-353.

Luddy, F., Barfield, R., Herb, S., Magidman, P. and Riemen Schneider, R. 1963. Pancreatic Lipase Hydrolysis of Triglycerides by a Semi micro Technique. *J Am Oil Chem Soc*, 41: 693-696.

Maes, S., Leventhal, H. and Ridder, D.T.D. 1996. Coping with chronic diseases. In Zeidner, M., and Endler, N. (Eds.), Handbook of Coping: Theory, Research and Applications, *Wiley,* New York, pp. 221-251.

Martin, A. 2001. Apports Nutritionnels Conseillés pour la population française, *Afssa, Cnerna-CNRS, Ed.Tec&Doc.*

Médale, F., Blanc, D. and Kaushik, S. 1991. Studies on the nutrition of Siberian sturgeon, *Acipenser baeri.* 2. Utilization of dietary non-protein energy by sturgeon. *Aquaculture,* 93: 143-154.

Médale, F., Lefèvre, F. and Corraze, G. 2003. Nutritional and dietary quality of fish: flesh constituents and factors of variation. *Cah Nutr Diét.* 38 : 37-44.

Mickleborough, T.D. and Rundell, K.W. 2005. Dietary polysaturated fatty acids in asthmaand -exerciseinduced -bronchoconstriction. *European Journal of Clinical Nutrition*, 59-:13351346.

Moyle, P., J. Cech. 2000. Fishes: An Introduction to Ichthyology - fourth edition. Upper Saddle River, NJ: Prentice-Hall.

Mozaffarian, D. and Rimm, E. 2006. Fish intake, contaminants, and human health: evaluating the risks and the benefits. *JAMA,* 296: 1885-1899.

Mozaffarian, D., Katan, M., Ascherio, A., Stampfer, M. and Willett, W. 2006. Transfatty acids and cardiovascular disease. *N Engl J Med,* 354: 1601-1613.

Nanton, D.A., Lall, S.P. and McNiven, M.A. 2001. Effects of dietary lipidlevel on liver and muscle lipid deposition in juvenile haddock, *Melanogrammusaeglefinus* L. *Aquacult. Res.* 32: 225-234.

Nanton, D., Lall, S., Ross, N. and McNiven, M. 2003. Effect of dietary lipid level on fatty acid β-oxidation and lipid composition in various tissues of haddock, Melanogrammus aeglefinus L. *Comp Biochem Physiol.* 135(B): 95-108.

Neddelman, P., Raz, A. and Minkes, M. 1979. Triene prostaglandins: protacylin and thromboxane synthesis and unique biological properties. *Proc Natl Acad Sci USA*, 76: 944-948.

Nelson, J. 1994. Fishes of the World (3rd edit), New York: Wiley & Sons.600p.

Öksüz, A., Özyılmaz, A. and Küver, Ş. 2011. Fatty acid composition and mineral content of Upeneus moluccensis and Mullus surmuletus. *Turkish Journal of Fisheries and Aquatic Sciences, 11*(1).

Olsen, Y. 1998. Lipids and essential fatty acids in aquatic food webs. In: Arts MT, Wainman BC (Eds) *Lipids in freshwater ecosystems*, p. 161-202.

Osman, H., Suriah, A.R. and E.C. Law. 2001. Fatty acid composition and cholesterol content of selected marine fish in Malaysian waters. Food Chemistry, 73: 55-60.

Ouannes-Ghorbel, A. 1996. Contribution à l'étude biologique des Labridés (Poissons Téléostéens Perciformes) des côtes de Sfax.*Rapp. DEA, Uni. Tunis II.* 118 p.

Ouannes-Ghorbel, A., Bradai, M.N. and Bouain, A. 2002. Reproduction period and sexual maturity of *Symphodus (Crenilabrus) tinca* (Labridae), from the Sfax coasts .*Cybium: 89-92.*

Ouannes-Ghorbel, A. 2003. Ecobiological study of Labridae (Fishes - Teleosts) from the southern coasts of Tunisia. *Thesis.* Fac. Sci. Sfax, 206p.

Ouannes-Ghorbel, A. and Bouain, A. 2006. The diet of the peacock wrasse, *Symphodus (Crenilabrus) tinca* (Labridae), in the southern coast of Tunisia.

Ownby, R.L., Crocco, E., Acevedo A., John V. and Loewenstein, D. 2006. Depression and risk for Alzheimer disease: systematic review, meta-analysis, and metaregression analysis.*Arch.Gen.Psychiatry*; 63(5):530-8.

Pallaoro A. and Jardas I. 2003. Some biological parameters of the peacock wrasse, *Symphodus (Crenilabrus) tinca* (L.1758) (Pisces: Labridae) from the Middle Eastern Adriatic(Croatian coast). *Sci. Mar.* 67(1): 33-41.

Pascal, J.C. and Ackman, R.G. 1976. Origin of di-isobutyl phthalate, a contaminant mimicking nonadecenoic acid in fatty acids of egg membrane lipids. *Comparative Biochemistry and Physiology Part B: Comparative Biochemistry*, *53*(1), pp.111-113.

Passi, S., Cataudella, S., Di Marco, P., De Simone, F., Rastrelli, L. 2002. Fatty Acid Composition and Antioxidant Levels in Muscle Tissue of Different Mediterranean Marine Species of Fish and Shellfish.*J. Agr. Food Chem.* 50: 7314-7322.

Peet, M., Murphy, B., Shay, J. and Horrobin, D. 1998. Depletion of omega-3 fatty acid levels in red blood cell membranes of depressive patients. *Biol Psychiatry,* 43: 315-319.

Puri, B., Ross, B. and Treasaden, I. 2008. Increased levels of ethane, a non-invasive, quantitative, direct marker of n-3 lipid peroxidation, in the breath of patients with schizophrenia. *Prog Neuropsychopharmacol Biol Psychiatry,* 32: 858-862.

Quignard, J.P. 1966. Recherches sur les Labridae (Poissons Téléostéens Perciformes) des côtes européennes. Systématique et biologie (Pisces, Teleosts, Perciformes) from the European coasts. Systematics and biology. Ed. *Cause et Castelnau. Montpellier*, 247p.

Quignrad, J.1978. Introduction à l'ichtyologie méditerranéenne: aspect général du peuplement. *Bull Off Natl Pêche Tunisie*, 2: 3-21

Quignard JP, Zaouali J. 1980. Les lagunes peri-medterraneenne. Bibliographie ichthyologique annotée. Premiere partie: Les etangs Frangais de Canet a Thau. Bull Off Nat Pech Tunis 4:293-360.

Quignard, J. and Pras, A. 1986. Labridae. In: Fishes of the North-Eastern Atlantic and the Mediterranean (Whitehead P.J.P., Bauchot M.-L., Hureau J.-C., Nielson J. & E. Tortonese,eds), pp. 919-942. Paris: *UNESCO*.

Quignard, J. and Tomasini, J. 2000. Mediterranean fish biodiversity. *Biol. Mar. Medit*, 7: 1-66.

Rasmussen, R. and Ostenfeld, T. 2000. Influence of growth rate on white muscle dynamics in rainbow trout and brook trout. *J Fish Biol.* 56: 1548-1552.

Rasmussen, R., Ostenfeld, T. and McLean, E. 2000. Growth and feed utilisation of rainbow trout subjected to changes in feed lipid concentrations. *Aquac. Int.*8: 531-542.

Regost, C., Arzel, J., Cardinal, M., Robin, J., Laroche, M. and Kaushik, S. 2001. Dietary lipid level, hepatic lipogenesis and flesh quality in turbot (*Psettamaxima)*. *Aquaculture,* 193: 291-309.

Reinitz, G. 1983. Relative effect of age, diet and feeding rate on the body, composition of young rainbow trout (*Salmo gairdneri*). *Aquaculture* 35: 19-27.

Rivaton J., Foumanoir P., Bourre.T P. and M. Kulbicki. 1989. Catalog des poissons de Nouvelle-Calédonie. *Catalogs Sci. Mer, ORSTOM Nouméa,* 2: 170 p.

Ross R.M., Losey G.S. and Diamond D.M. 1983. Sex change in a coral-Reef fish: dependence of stimulation and inhibitionon relative size. *Science,* 221: 574-575

Sargent, J., Henderson, R. and Tocher, D. 1989. The lipids. In: Fish Nutrition (Eds: Halver, J.E.). *Academic Press, INC,* pp. 153-218.

Sargent, J., Bell, G., McEvoy, L., Toucher, D. and Estevez, A. 1999. Recent developments in the essential fatty acid nutrition of fish. *Aquaculture,* 177: 191-199.

Sargent, J., Bell, J., Bell, M., Henderson, R. and Tocher, D. 1995. Requirement criteria for essential fatty acids. *Journal of Applied Ichthyology-Zeitschrift Fur AngewandteIchthyologie* 11: 183-198.

Sargent, J.R., McEvoy, L.A. and Bell, J.G. 1997. Requirements, presentation and sources of polyunsaturated fatty acids in marine fish larval feeds. *Aquaculture, 155*(1-4), pp.117-127.

Schmitz, G. and Ecker, J. 2008. The opposing effects of n-3 and n-6 fatty acids. *Progress in Lipid Research,* 47: 147-155.

Seloudre, P. and C. Chauvet. 1986. Preliminary observations on the ichthyofauna of a superficial meadow, the Posidonia meadow of Racou (Golfe du Lion). Rapp. Comm. Int. mer Médit, 30: 224.

Shearer, K. 1994. Factors affecting the proximate composition of cultured fishes with emphasis on salmonids. *Aquaculture* 119: 63-88.

Shearer, K., Silverstein, J. and Dickhoff, W. 1997. Control of growth and adiposity of juvenile Chinook salmon (*Oncorhynchus tshawytscha*). *Aquaculture* 157: 311-323.

Sigurgisladottir, S., Lall, S.P., Parrish, C.C. and Ackman, R.G. 1992. Cholestane as a digestibility marker in the absorption of polyunsaturated fatty acid ethyl esters in Atlantic salmon. *Lipids* 27, 418-424.

Soljan, T. 1930. Nest of Adriatic fishes Lipp (*Crenilabrus ocellatus Forsk.*) - *Z. Morphol. ecol Animals* 17, p. 145-153.

Soljan, T. 1931. Parental care by nest-building at Crenilabrus quinquema culata Risso, an Adriatic wrasse. - *Z. Morphol. ecol Animals* 20, p. 132-135

Sonntag, N.O.V.b. 1979. Structure and composition of fats and oils. In: Swern D. (Ed.), Bailey's industrial oil and fat products, volume 1. Fourth edition. *Wiley-Interscience: New-York,* 1-98.

Stephenson, C. 2004. Fish oil and inflammatory disease: Is asthma the next target for n-3 fatty acid supplements, *Nutrition Reviews*, 62(12):486489-.

Stickland, N.C. 1983. Growth and development of muscle fibers in the rainbow trout (*Salmo gairdneri*). *J.Anat.* 7, 323-333.

Stoner, A. 1980. Feeding ecology of *Lagodon rhomboides* (Pisces: Sparidae): Variation and functional responses. Sympatric sparid fishes from sea-grass meadows. *Fish Bull*, 78: 337-352.

Stoner, A. and Lingviston, R. 1984. Ontogenetic patterns in diet and feeding morphology in sympatric sparid fishes from sea-grass meadows. *Copeia*, 1984: 174-178.

Takeuchi, T., Watanabe, T. and Ogino, C. 1978. Supplementary effects of lipids in a high protein diet for rainbow trout. *Bull Japan Soc Sci Fish.* 44: 677-681.

Takeuchi, T., Watanabe, T. and Ogino, C. 1979. Digestibility of hydrogenated fish oils in carp and rainbow trout. *Bull Japan Soc Sci Fish,* 45: 1517-1519.

Tavazzi, L., Maggioni, A., Marchioli, R., Barlera, S., Franzosi, M., Latini, R., Lucci, D., Nicolosi,G., Porcu, M. and Tognoni, G. 2008. Effect of n-3 polyunsaturated fatty acids in patients withchronic heart failure (the GISSI-HF trial): a randomised, double-blind, placebo-controlled trial, *Lancet,* 372: 1223-1230.

Thresher, R. 1983. Habitat effect on reproductive success in the coral reef fish, *Acanthochromis Polyacanthus* (Pomacentriadea). *Ecology,* 64:1184-1199.

Tocher, D. 2003. Metabolism and functions of lipids and fatty acids in teleost fish. *Reviews in Fisheries Science,* 11: 107-184.

Torstensen, B.E., Lie, Ø. and Frøyland, L. 2000. Lipid Metabolism and Tissue Composition in Atlantic Salmon (*Salmo salar L.*)-Effects of Capelin Oil, Palm Oil, and Oleic Acid-Enriched Sunflower Oil as Dietary Lipid Sources. *Lipids* 35, 653-664.

Tortonese E. 1975. Osteichthyes, Pesci Ossei. In: Fauna d'Italia. Calderini Ed. Bologna, 6: 374-376.

Turon, F., Bachain, P., Caro, Y., Pina, M. and Graille, J. 2002. A direct method for regiospecific analysis of TAG using α-MAG. *Lipids,* 37: 817-821.

Uchiyama, H. and S, Ehira. 1974. Relation between freshness and acid-soluble nucleotides in aseptic cod and yellowtail muscles during ice storage. *Bull. Tokai Reg. Fish. Lab. 78,* 23-31.

Van Anholt, R., Spanings, E., Koven, W., Nixon, O. and Bonga, S. 2004. Arachidonic acid reduces the stress response of gilthead sea bream *Sparus aurata* L. *Journal of Experimental Biology,* 207: 3419-3430.

Van der Kooy, K., Van Hout, H., Marwijk, H. 2007. *Depression and the risk for cardiovascular diseases: Systematic review and meta-analysis. Int J Geriatr Psychiatry; 22:613-26.*

Veylon, R. 1977. La mer, source de médicaments. *La nouvelle presse médicale*, 6: 2353-2356.

Walton, M.J. and Cowey, C.B. 1982. Aspects of intermediary metabolism in salmonid fish. *Comp. Biochem. Physiol.* 71(B), 59-79.

Ware, D.M. 1972. Predation by rainbow trout (*Salmo gairdneri*): the influence of hunger, prey density and prey size. *J Fish Res Board Can*, 29: 1193-1201. .

Wassall, S.R., Brzustowicz, M.R., Shaikh, S.R., Cherezov, V., Caffrey, M. and Stillwell, W. 2004. Order from disorder, corralling cholesterol with chaotic lipids: The role of polyunsaturated lipids in membrane raft formation. *Chemistry and physics of lipids*, *132*(1), pp.79-88.

Watanabe, T. 1982. Lipid nutrition in fish.*Comp. Biochem. Physio.* 73: 3-15.

Watkins, B., Lippmann, H., Le Bouteiller, L., Li, Y. and Seifert, M. 2001. Bioactive fatty acids: role in bone biology and bone cell function. *Prog Lipid Res*, 10: 125-148.

Weatherley, A.H. and Gill, H.S. 1983. Relative growth of tissues at differentsomatic growth rates in rainbow trout *Salmo gairdneri* Richardson. *J. Fish Biol.* 23,43-60.

Zhou, S., Ackman R. and Morrison, C. 1996. Adipocytes and Lipid distribution in the muscle tissue of Atlantic salmon (*Salmo salar). Can J Fish AquatSci,.*53: 326-332.

Zivkovic, A., German, J. and Sanyal, A. 2007. Comparative review of diets for the metabolic syndrome: implications for nonalcoholic fatty liver disease. *Am J Clin Nutr,* 86: 285-300.

Appendices

<u>**Appendix1:**</u>

Main naturally occurring saturated fatty acids (International Union of Pureand Applied Chemistry and International Union of Biochemistry Commission on Biochemical Nomenclature,1978; Kramer et al., 1998).

Standard nomenclature	Systematic acid name	Acid commonname	Structurechemical	Omega nomenclature
C1: 0	Methanoic	Formic	CHOOH	C1:0
C2 :0	Ethanoic	Acetic	CH3COOH	C2:0
C3 :0	Propanoic	Propionic	CH3CH2COOH	C3:0
C4 :0	Butanoic	Butyric	CH3(CH2)2COOH	C4:0
C5 :0	3-Methylbutanoic acid	Valeric	CH3(CH2)3COOH	C5:0
C5:0iso	Pentanoic	Isovaleric	CH3CHCH3CH2COOH	C5:0iso
C6 :0	hexanoic	Caproic acid	CH3(CH2)4COOH	C6:0
C7 :0	Heptanoic	Enanthic	CH3(CH2)5COOH	C7:0
C8 :0	Octanoic	Caprylic	CH3(CH2)6COOH	C8:0
C9 :0	Nonanoic	Pelargonic	CH3(CH2)7COOH	C9:0
C10 :0	Decanoic	Caprique	CH3(CH2)8COOH	C10:0
C12 :0	Dodecanoic	Laurique	CH3(CH2)10COOH	C10:0
C14 :0	Tetradecanoic	Myristic	CH3(CH2)12COOH	C14:0
C16 :0	Hexadecanoic	Palmitic	CH3(CH2)14COOH	C16:0
C18 :0	Octadecanoic	Stearic	CH3(CH2)16COOH	C18:0
C20 :0	icosanoic1	Peanut	CH3(CH2)18COOH	C20:0
C22 :0	Docosanoic	Behenics	CH3(CH2)20COOH	C22:0
C24 :0	Tetracosanoic	Iignoceric	CH3(CH2)22COOH	C24:0
C26 :0	Hexacosanoic	Cerotic	CH3(CH2)24COOH	C26:0
C28 :0	Octacosanoic	Montanique	CH3(CH2)26COOH	C28:0
C30 :0	Tricontanoic	melissistic	CH3(CH2)28COOH	C30:0

Main monounsaturated fatty acids found in nature (International Union of Pureand Applied Chemistry and International Union of Biochemistry Commission on Biochemical Nomenclature, 1978; Kramer et al., 1998).

Standard nomenclature	Systematic acid naming	Acid commonname	Structurechemical	Omega nomenclature
C12:1(9)	cis-9-dodecenoic acid	lauroleic	$CH_3CH_2CH=CH(CH_2)_7COOH$	C12:1w-3
C14:1(9)	tetradecenoic	Myristoleic	$CH_3(CH_2)_3CH=CH(CH_2)_7COOH$	C14:1w-5
C16:1(9)	cis-9-hexadecenoic acid	Palmitoleic	$CH_3(CH_2)_5CH=CH(CH_2)_7COOH$	C16:1w-7
C18:1(trans 6)	trans-6-octadecenoic acid	Petroselaidic	$CH_3(CH_2)_{10}CH=CH(CH_2)_4COOH$	C18:1w-12
C18:1(9)	cis-9-octadecenoic acid	Oleic	$CH_3(CH_2)_7CH=CH(CH_2)_7COOH$	C18:1w-9
C18:1(trans 9)	trans-9-octadecenoic acid	Elaidic	$CH_3(CH_2)_7CH=CH(CH_2)_7COOH$	C18:1w-9
C18:1(11)	cis-11-octadecenoic acid	Vaccenic	$CH_3(CH_2)_5CH=CH(CH_2)_9COOH$	C18:1w-7
C18:1(trans 11)	trans-11-octadecenoic acid	transvaccenic	$CH_3(CH_2)_5CH=CH(CH_2)_9COOH$	C18:1w-7
C20:1(9)	cis-9-icosenoic(1)	Gadoleic	$CH_3(CH_2)_9CH=CH(CH_2)_7COOH$	C20:1w-11
C22:1(11)	cis-11-docosenoic acid	Ketoleic	$CH_3(CH_2)_9CH=CH(CH_2)_9COOH$	C22:1w-11
C22:1(13)	cis-13-docosenoic acid	Erucic	$CH_3(CH_2)_7CH=CH(CH_2)_{11}COOH$	C22:1w-9
C24:1(15)	cis-15-tetracosenoic acid	Selacholeic	$CH_3(CH_2)_7CH=CH(CH_2)_{13}COOH$	C24:1w-9

Main naturally occurring polyunsaturated fatty acids (International Union of Pureand AppliedChemistry and International Union of Biochemistry Commission on Biochemical Nomenclature, 1978; Kramer et al., 1998).

Standard nomenclature	Acid namesystem	Common acid name	Structural-chemical	Nomenclatureomega
C18:2(9, 12)	cis-9,12-octadecadienoic acid	linoleic	$CH_3(CH_2)_4CH=CHCH_2CH=CH(CH_2)_7COOH$	**C18:2ω 6**
C18 : 2(9,trans11)	cis,trans-9,11-octadecadienoic acid	linoleic conjugate-9, trans-11 ouruménique	$CH_3(CH_2)_5CH=CHCH=CH(CH_2)_7COOH$	C18:2ω 7
C18:2(trans 10,12)	trans,cis-10,12-octadecadienoic acid	linoleic conjugué trans-10, cis-12	$CH_3(CH_2)_4CH=CHCH=CH(CH_2)_8COOH$	C18:2ω 6
C18 : 3(6,9,12)	cis,cis,cis-6,9,12-octadecatrienoic acid	g-linolenic	$CH_3(CH_2)_3(CH_2CH=CH)_3(CH_2)_4COOH$	C18:3 ω6
C18 : 3(9,12, 15)	cis,cis,cis-9,12,15-octadecatrienoic acid	a-linolenic	$CH_3(CH_2CH=CH)_3(CH_2)_7COOH$	C18:3ω 3
C18 : 3(9,trans 11,trans13)	cis,trans,trans-9,11,13-octadecatrienoic	a-eleostearic	$CH_3(CH_2)_3(CH=CH)_3(CH_2)_7COOH$	C18:3ω 5
C20 : 4(5,8,11,14)	cis,cis,cis,cis-5,8,11,14-icosatetraenoic(1)	Arachidonic	$CH_3(CH_2)_4(CH=CHCH_2)_3CH=CH(CH_2)_3COOH$	C20:4ω 6
C20 : 5(5,8,11,14,17)	cis,cis,cis,cis,cis-5, 8, 11, 14,17-icosapentaenoic acid1	EPA	$CH_3CH_2(CH=CHCH_2)_4CH=CH(CH_2)_3COOH$	C20:5ω 3
C22 : 5(4,8,12,15,19)	cis,cis,cis,cis,cis-4, 8,12, 15,19-docosapentaenoic acid	Cuplanodonic	$CH_3CH_2CH=CH(CH_2)_2CH=CHCH_2(CH=CH(CH_2)_2)_3COOH$	C22:5ω 3
C22 : 6(4,7,10,13,16,19)	cis,cis,cis,cis,cis,cis-4,7,10,13,16,19-	DHA	$CH_3CH_2(CH=CHCH_2)_5CH=CH(CH_2)_2COOH$	C22:6ω 3

docosahexaen oic

<u>**Appendix2**</u>

Percentages of fatty acids present in the total lipids of the flesh, liver and gonads of *Symphodus tinca,* living in the marine environment.

	Chair		Livers		gonads	
	♀	♂	♀	♂	♀	♂
C14:0	1,74%	2,11%	3,80%	5,75%	1,74%	2,11%
C16:0	23,17%	26,18%	29,00%	26,37%	23,17%	26,18%
C17:0	1,14%	0,37%	2,18%	0,79%	0,34%	0,46%
C18:0	7,84%	6,43%	7,20%	7,13%	7,84%	6,43%
C16:1n-7	2,76%	4,76%	5,09%	10,03%	2,76%	4,76%
C18:1n-9	7,58%	10,99%	7,96%	14,75%	7,58%	10,99%
C18:2 n-6	1,28%	2,16%	2,59%	3,20%	0,97%	2,76%
C18:3 n-6	0,64%	0,33%	0,14%	0,32%	1,36%	0,26%
C18:3n-3	0,44%	1,69%	0,55%	2,49%	0,40%	1,22%
C20:2 n-6	0,82%	0,87%	0,14%	0,80%	1,22%	1,29%
C20:3 n-6	0,35%	0,37%	0,07%	0,38%	0,54%	0,42%
C20:4n-6	14,98%	11,57%	10,39%	8,39%	14,98%	11,57%
C20:5n-3	11,89%	11,74%	9,36%	11,18%	11,89%	11,74%
C22:5n-3	4,17%	3,26%	1,69%	1,29%	4,17%	3,26%
C22:6 n-3	23,30	15,66	17,30	7,06	23,30	15,66

Percentages of fatty acids present in the total lipids of the flesh, liver and gonads of *Symphodus tinca,* living in lagoons.

	Chair		Livers		gonads	
	♀	♂	♀	♂	♀	♂
C14:0	4,31%	3,05%	4,66%	8,84%	23,70%	21,93%
C16:0	23,67%	21,81%	24,35%	30,69%	4,77%	1,70%
C17:0	0,5%	0,42%	0,93%	0,95%	0,51%	0,63%
C18:0	5,97%	5,81%	6,94%	5,10%	5,01%	4,39%
C16:1n-7	5,78%	6,90%	12,73%	13,30%	10,05%	5,29%
C18:1n-9	12,50%	10,17%	11,25%	11,08%	18,20%	15,05%
C18:2 n-6	1,71%	0,42%	2,81%	3,20%	2,60%	1,55%
C18:3 n-6	0,35%	0,92%	0,37%	0,32%	0,28%	0,68%
C18:3n-3	0,49%	0000%	0,82%	2,44%	0,72%	0,91%
C20:2 n-6	1,01%	0,64%	1,72%	0,84%	0,82%	0,62%
C20:3 n-6	0,39%	0000%	0,26%	0,09%	0,02%	0,10%
C20:4n-6	10,28%	10,80%	7,30%	6,50%	10,39%	12,65%
C20:5n-3	11,88%	8,30%	12,90%	8,04%	12,24%	13,82%
C22:5n-3	2,19%	1,87%	2,06%	1,12%	1,85%	7,16%
C22:6 n-3	17,6%	26,83%	9,56%	7,77%	13,20%	21,38%

Percentages of fatty acids present in the total lipids of the flesh, liver and gonads of *Symphodus tinca,* found on island coasts.

	Chair		Livers		Gonads	
	♀	♂	♀	♂	♀	♂
C14:0	6,09%	5,39%	7,35%	5,88%	5,00%	3,28%
C16:0	28,02%	24,76%	26,55%	25,70%	26,95%	26,48%
C17:0	0,35%	0,30%	0,34%	0,45%	0,32%	0,53%
C18:0	6,67%	6,54%	5,63%	4,99%	5,46%	7,27%
C16:1n-7	7,75%	9,20%	10,48%	10,52%	6,52%	5,50%
C18:1n-9	18,43%	21,12%	18,27%	20,18%	15,77%	20,96%
C18:2 n-6	2,84%	2,18%	2,01%	1,39%	1,75%	2,61%
C18:3 n-6	0,50%	0,45%	0,48%	0,37%	0,52%	0,36%
C18:3n-3	0,51%	0,75%	0,13%	1,21%	1,07%	0,58%
C20:2 n-6	0,78%	0,71%	0,17%	0,96%	0,75%	0,74%
C20:3 n-6	0,69%	0,54%	0,04%	0,45%	0,22%	0,4%
C20:4n-6	6,77%	8,00%	6,58%	8,55%	8,47%	10,03%
C20:5n-3	3,48%	1,54%	11,27%	10,22%	10,03%	7,94%
C22:5n-3	6,41%	4,05%	3,06%	2,48%	3,74%	2,60%
C22:6 n-3	11,67%	9,88%	6,83%	5,36%	13,12%	9,15%

<u>**Appendix3**</u>

Variation in the percentage of C14:0, C16:0 and C18:0 in the flesh of male and female *symphodus tina* at the marine station (month of March).

%inAGT	Female fillets	Male fillets
C14:0	1,74%	2,11%
C16:0	23,17%	26,18%
C18:0	7,84%	6,43%

Variation in the percentage of C14:0, C16:0 and C18:0 in the flesh of male and female *symphodustina* at the lagoon station (month of March).

%inAGT	Female fillets	Male fillets
C14:0	4,31%	3,05%
C16:0	23,67%	21,81%
C18:0	5,97%	5,81%

Variation in the percentage of C14:0, C16:0 and C18:0 in the flesh of male and female *symphodus tina* at the island station.

%inAGT	Female fillets	Male fillets
C14:0	6,09%	5,39%
C16:0	28,02%	24,76%
C18:0	6,67%	6,54%

Chromatogramme des AGT de la chair d'un individu mâle(milieu marin)

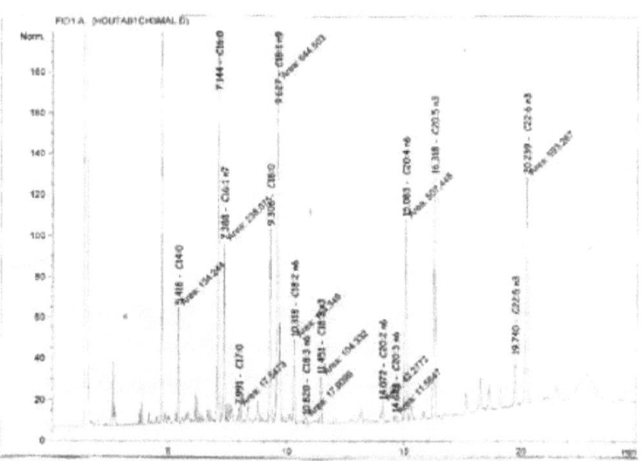

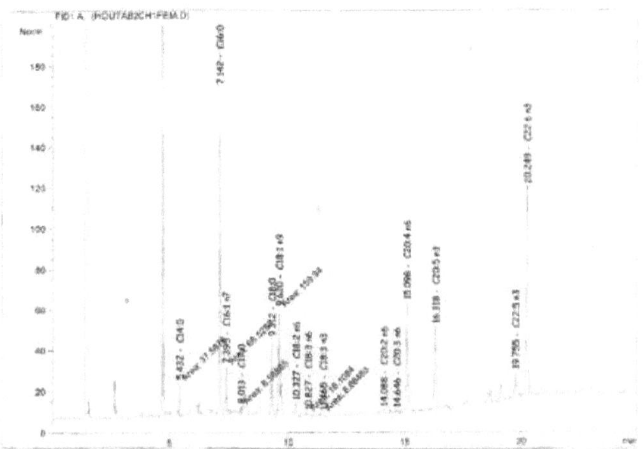

Chromatogramme des AGT de la chair d'un individu mâle (milieu lagunaire)

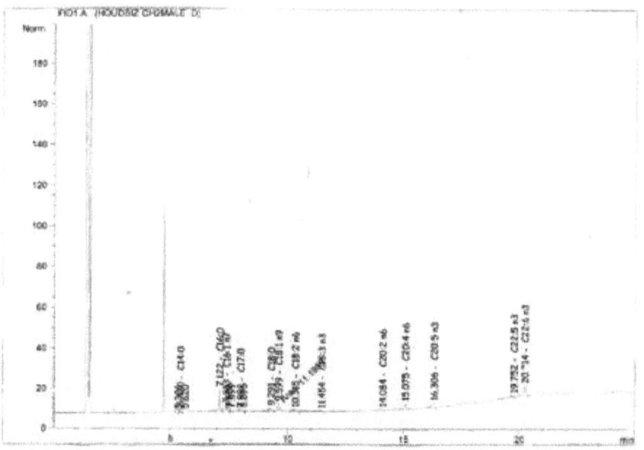

Chromatogramme des AGT de la chair d'un individu femelle(milieu lagunaire)

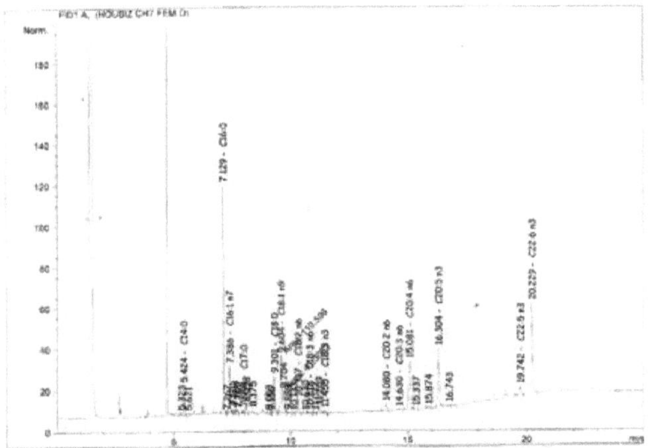

Chromatogramme des AGT de la chair d'un individu mâle (milieu insulaire)

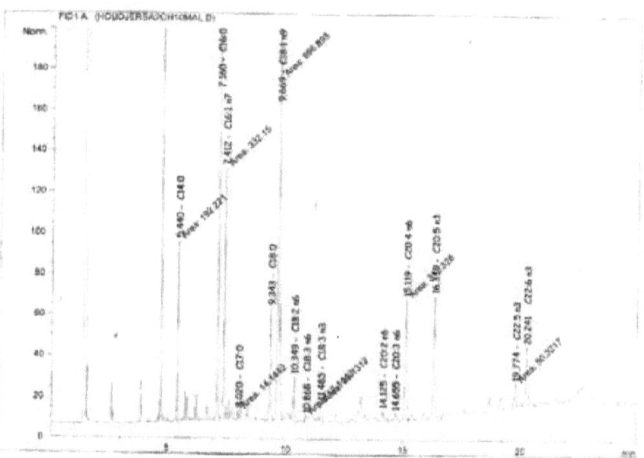

Chromatogramme des AGT de la chair d'un individu femelle(milieu insulaire)

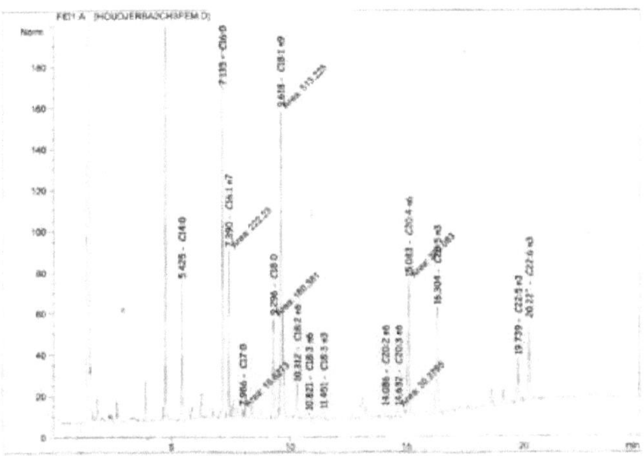

Chromatogramme des AGT hépatiques d'un individu mâle (milieu marin)

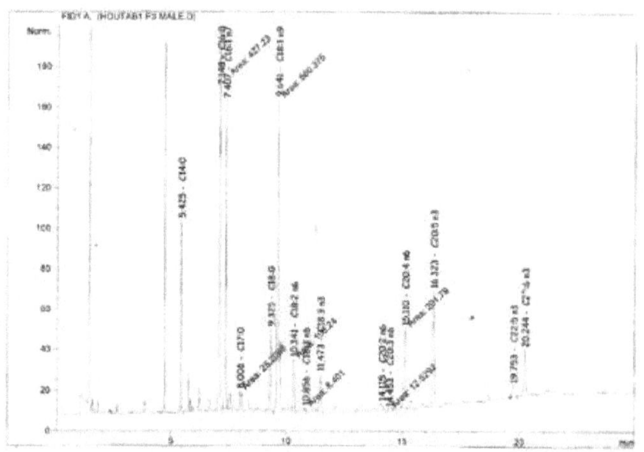

Chromatogramme des AGT hépatique d'un individu femelle (milieu marin)

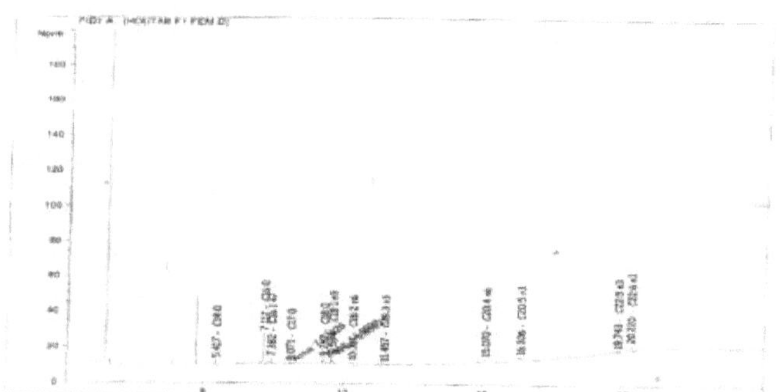

Chromatogramme des AGT hépatiques d'un individu mâle (milieu lagunaire)

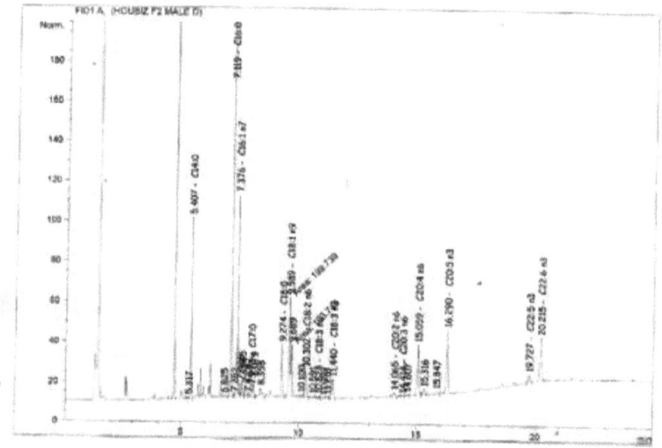

Chromatogramme des AGT hépatiques d'un individu femelle(milieu lagunaire)

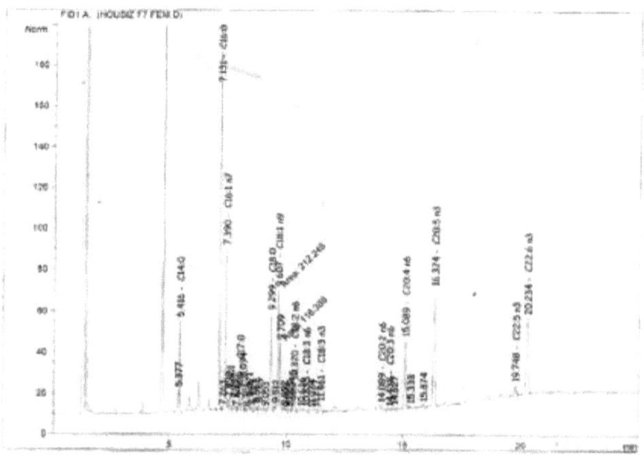

Chromatogramme des AGT hépatiques d'un individu mâle (milieu insulaire)

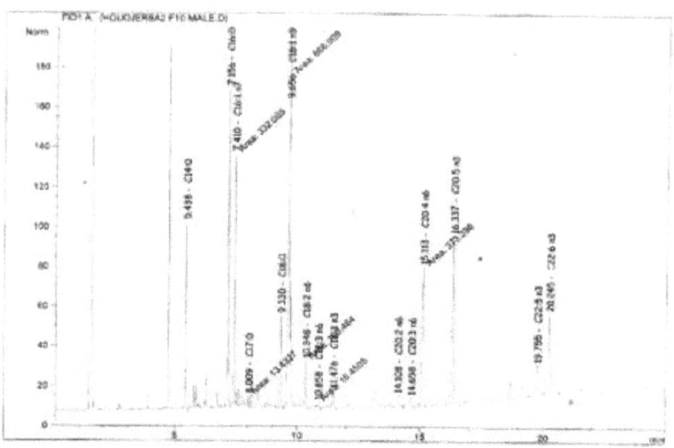

Chromatogramme des AGT hépatiques d'un individu femelle(milieu insulaire)

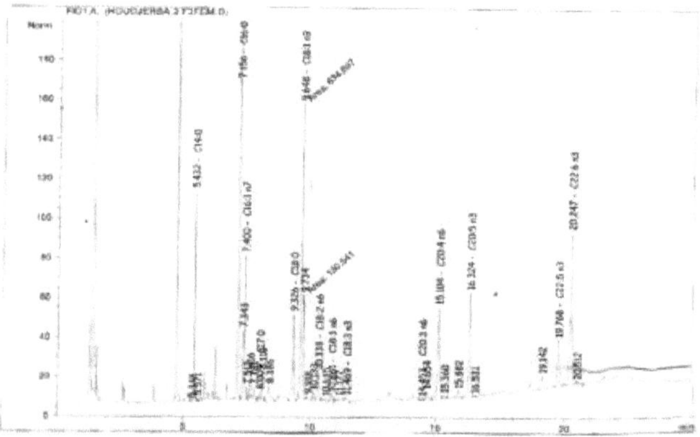

Chromatogramme des AGT gonadiques d'un individu mâle (milieu marin)

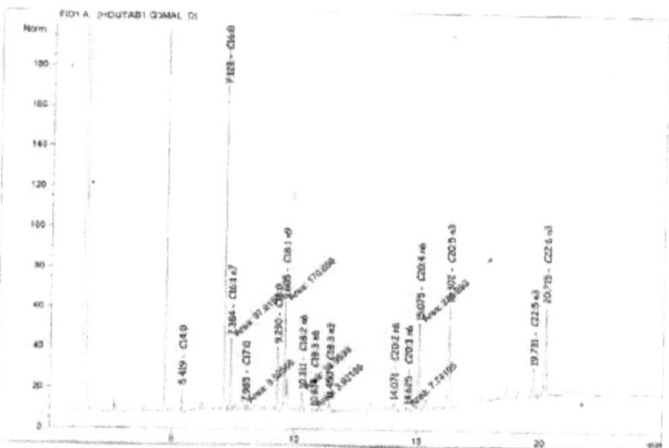

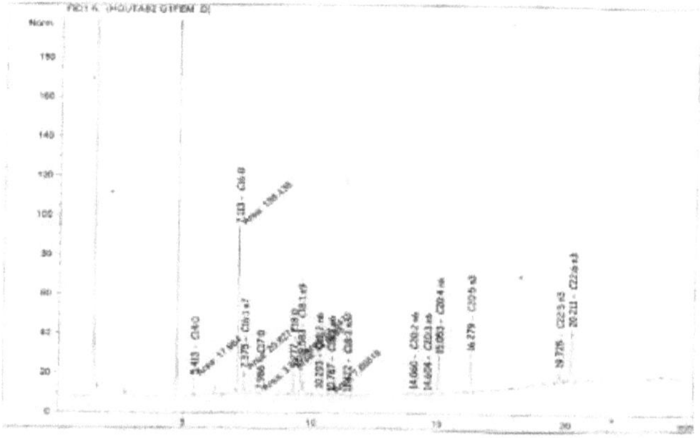

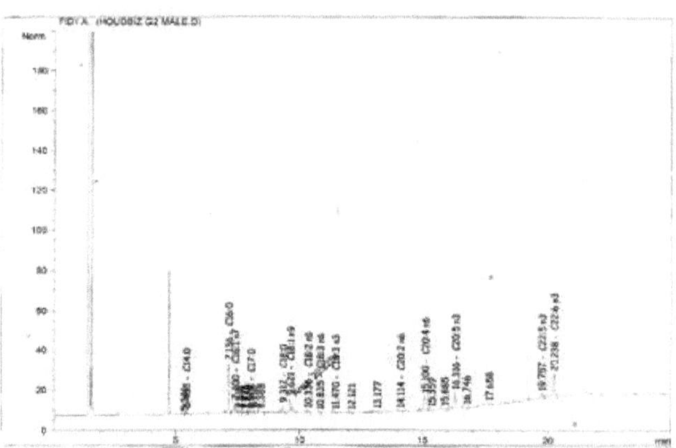

Chromatogramme des AGT gonadiques d'un individu femelle (milieu lagunaire)

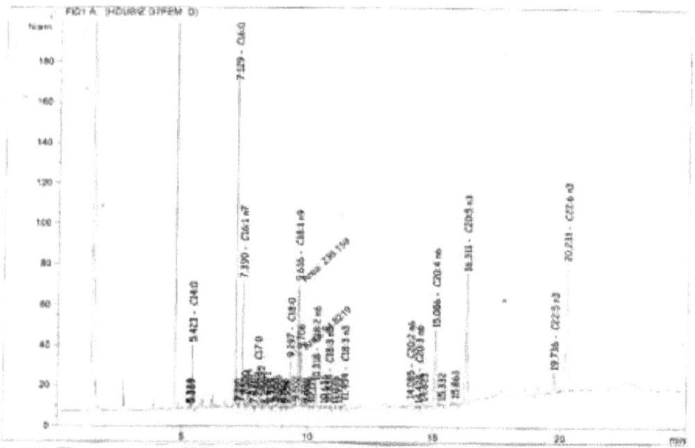

Chromatogramme des AGT gonadiques d'un individu mâle (milieu insulaire)

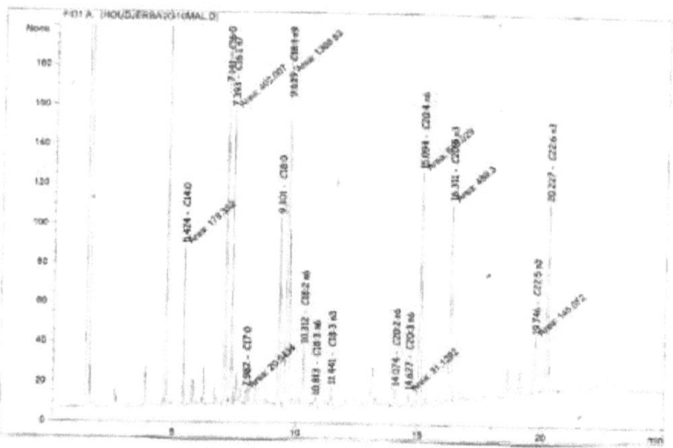

Chromatogramme des AGT gonadiques d'un individu femelle (milieu insulaire)

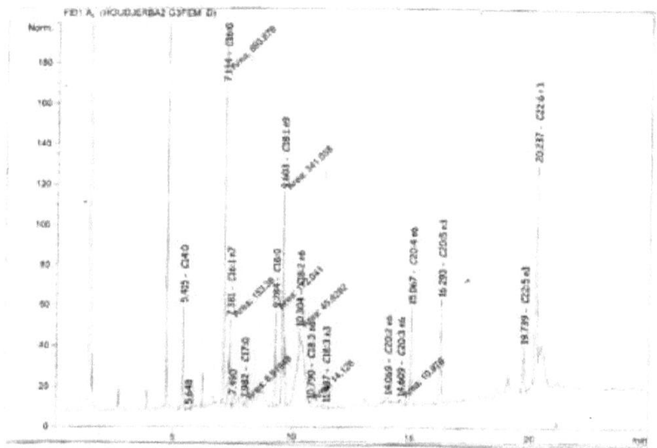

I want morebooks!

Buy your books fast and straightforward online - at one of world's fastest growing online book stores! Environmentally sound due to Print-on-Demand technologies.

Buy your books online at
www.morebooks.shop

Kaufen Sie Ihre Bücher schnell und unkompliziert online – auf einer der am schnellsten wachsenden Buchhandelsplattformen weltweit! Dank Print-On-Demand umwelt- und ressourcenschonend produziert.

Bücher schneller online kaufen
www.morebooks.shop

info@omniscriptum.com
www.omniscriptum.com

MIX
Papier aus verantwortungsvollen Quellen
Paper from responsible sources
FSC® C105338

Printed by Books on Demand GmbH, Norderstedt / Germany